Fracking

Recent Titles in the

CONTEMPORARY WORLD ISSUES

Series

Military Robots and Drones: A Reference Handbook
Paul J. Springer

Marijuana: A Reference Handbook
David E. Newton

Religious Nationalism: A Reference Handbook
Atalia Omer and Jason A. Springs

The Rising Costs of Higher Education: A Reference Handbook
John R. Thelin

Vaccination Controversies: A Reference Handbook
David E. Newton

The Animal Experimentation Debate: A Reference Handbook
David E. Newton

Steroids and Doping in Sports: A Reference Handbook
David E. Newton

Internet Censorship: A Reference Handbook
Bernadette H. Schell

School Violence: A Reference Handbook, Second Edition
Laura L. Finley

GMO Food: A Reference Handbook
David E. Newton

Wind Energy: A Reference Handbook
David E. Newton

Profiling and Criminal Justice in America: A Reference Handbook, Second Edition
Jeff Bumgarner

Books in the **Contemporary World Issues** series address vital issues in today's society such as genetic engineering, pollution, and biodiversity. Written by professional writers, scholars, and nonacademic experts, these books are authoritative, clearly written, up to date, and objective. They provide a good starting point for research by high school and college students, scholars, and general readers as well as by legislators, businesspeople, activists, and others.

Each book, carefully organized and easy to use, contains an overview of the subject, a detailed chronology, biographical sketches, facts and data and/or documents and other primary source material, a forum of authoritative perspective essays, annotated lists of print and nonprint resources, and an index.

Readers of books in the *Contemporary World Issues* series will find the information they need in order to have a better understanding of the social, political, environmental, and economic issues facing the world today.

CONTEMPORARY WORLD ISSUES

Fracking

A REFERENCE HANDBOOK

David E. Newton

Santa Barbara, California • Denver, Colorado • Oxford, England

Library of Congress Cataloging-in-Publication Data

Newton, David E., author.
Fracking : a reference handbook / David E. Newton.
pages cm. — (Contemporary world issues)
Includes bibliographical references and index.
ISBN 978–1–61069–691–3 (hardback : acid-free paper) — ISBN 978–1–61069–692–0 (ebook) 1. Hydraulic fracturing—Environmental aspects. 2. Gas well drilling—Environmental aspects. 3. Hydraulic fracturing—Government policy—United States. I. Title.
TD195.G3N49 2015
622'.2—dc23 2014038749

ISBN: 978–1–61069–691–3
EISBN: 978–1–61069–692–0

19 18 17 16 15 1 2 3 4 5

This book is also available on the World Wide Web as an eBook. Visit www.abc-clio.com for details.

ABC-CLIO, LLC
130 Cremona Drive, P.O. Box 1911
Santa Barbara, California 93116-1911

This book is printed on acid-free paper ∞

Manufactured in the United States of America

Preface, xv

1 **Background and History, 3**

The Genesis of Fossil Fuels, 7

Oil and Gas Resources, 13
    Terminology, 18

Exploring for Oil and Gas, 20

Drilling Technology, 23

Petroleum Chemistry, 27

A Brief History of Petroleum Exploration, 30

The Petroleum Century, 34

The Natural Gas Century, 39

Hydraulic Fracturing, 43
    The History of Hydraulic Fracturing, 44
    Horizontal Drilling, 49

Conclusion, 51

References, 52

2 Problems, Controversies, and Solutions, 63

Peak Oil and Gas: Still a Looming Threat or Only a Distant Fantasy?, 63

The Global State of Affairs, 68

Energy Independence for the United States, 70

Benefits of Fracking, 76

Direct Economic Benefits, 76

Indirect Economic Benefits, 79

Environmental Benefits, 80

Opposition to Fracking, 82

Problems Associated with Hydraulic Fracturing, 84

Water Use, 84

Water Contamination, 87

Air Pollution, 92

Earthquakes, 96

Aesthetic and Related Disturbances, 100

Laws and Regulations, 101

Conclusion, 112

References, 113

3 Perspectives, 131

Introduction, 131

Separating Fact from Hype: Trudy E. Bell, 131

Why Fracking Is Beneficial: Bruce Everett, 136

The Urgent Need for Global Definition of Terms in the Reporting on Fracking and Seismic Activity: Gina Hagler, 140

Fracking and the Future of Fresh Water: Michael Pastorkovich, 145

Fracking by the Numbers: John Rumpler, 149

Exploring Alternative Uses for Fracking Water: Lana Straub, 154

Fracking Like It's Your Job: Why Worker Safety in the Fracking Industry Is So Important: Laura Walter, 158

Feeding the Fracking Workforce: How Worker Nutrition Supports Health, Safety and Productivity: Christopher Wanjek, 162

Food as Protection, 162

What Workers Need, and What Workers Get, 163

Return on Investment, 165

References, 166

4 **Profiles, 169**

Introduction, 169

American Gas Association, 170

American Petroleum Institute, 172

America's Natural Gas Alliance, 175

Chesapeake Climate Action Network, 176

Consumer Energy Alliance, 179

H. John Eastman (1905–1985), 181

Terry Engelder, 184

Environment America, 185

Food & Water Watch, 187

Ground Water Protection Council, 190

Erle P. Halliburton (1892–1957), 192

William Hart (1797–1865), 194

Independent Petroleum Association of America, 196

Anthony R. Ingraffea (1947–), 200

International Energy Agency, 202

Interstate Natural Gas Association of America, 206

Interstate Oil and Gas Compact Commission, 208

E. W. Marland (1874–1941), 211

Aubrey McClendon (1959–), 213

George Mitchell (1919–2013), 215

The National Association of Royalty Owners, 218

No Fracked Gas in Mass, 220

Edward A. L. Roberts (1829–1881), 223

The Endocrine Disruption Exchange, 226

Tom L. Ward (1959–), 228

Waterkeeper Alliance, 230

5 Data and Documents, 235

Introduction, 235

Data, 235

Table 5.1 Chemicals Present in 652 Different Products Used by Hydraulic Fracturing Companies, 236

Table 5.2 Discovered but Unproved Technically Recoverable Shale Gas and Oil Resources in the United States: Shale Oil Resources, 237

Table 5.3 Estimated Number of Fracking Wells in the United States, 238

Table 5.4 Water Used for Fracking, 239

Table 5.5 Estimated Air Pollution Produced from Early Stages of Fracking (Drilling and Well Completion) in 2012 (tons), 239

Table 5.6 Estimated Risked Shale Gas and Shale Oil Resources In-Place and Technically Recoverable in 41 Countries as Assessed in 2013, 240

Table 5.7 U.S. Crude Oil and Natural Gas Proved Reserves, 242

Documents, 243

Crude Oil Windfall Profit Tax Act of 1980, 243

Hydraulic Fracturing Exclusions, 246

Safe Drinking Water Act, Public Law 113-103, Section 300h(d)(1), 246

Clean Water Act, Public Law 95-217, Sections 1326, 1342, 247

Resource Conservation and Recovery Act (RCRA), Public Law 94-580, 6921(b), 248

Comprehensive Environmental Response, Compensation, and Liability Act (CERCLA; also known as the Superfund Act), Public Law 96-510, 9601, 250

Legal Environmental Assistance Foundation vs. U.S. EPA, 118 F.3d 1467 (1997), 251

Evaluation of Impacts to Underground Sources of Drinking Water by Hydraulic Fracturing of Coalbed Methane Reservoirs Study (2004), 252

ES-8 Did EPA Find Any Cases of Contaminated Drinking Water Wells Caused by Hydraulic Fracturing in CBM Wells?, 253

Testimony Submitted to the House Committee on Natural Resources Subcommittee on Energy and Mineral Resources Washington, DC, June 18, 2009, Prepared by the Interstate Oil and Gas Compact Commission on Behalf of the Nation's Oil and Gas Producing States, 254

Study of the Potential Impacts of Hydraulic Fracturing on Drinking Water Resources—Progress Report (2012), 256

Act 13, State of Pennsylvania (2012), 258

Robinson Township, et al. vs. Commonwealth of Pennsylvania, (J-127A-D-2012) (2013), 260

Fracturing Responsibility and Awareness of Chemicals Act S.1135, 113th Congress, 1st Session (2013), 263

Powder River Basin Resource Council, Wyoming Outdoor Council, Earthworks, and Center for Effective Government (Formerly OMB Watch) vs. Wyoming Oil and Gas Conservation Commission and Halliburton Energy Services, Inc., 2014 WY 37 (2014), 266

Norse Energy Crop. vs. Town of Dryden, et al. (2014), 268

Communication from the Commission to the European Parliament, the Council, the European Economic and Social Committee and the Committee of the Regions on the Exploration and Production of Hydrocarbons (Such as Shale Gas) Using High Volume Hydraulic Fracturing in the EU (2014), 270

6 Resources for Further Research, 275

Introduction, 275

Books, 275

Articles, 284

Reports, 296
Internet, 302

7 **CHRONOLOGY, 317**

Glossary, 327
Index, 333
About the Author, 339

Preface

For more than half a decade, U.S. presidential administrations have been calling on the nation to become "energy independent." In using these terms, presidents from Richard Nixon to Barack Obama have pointed out the advantages of having the United States capable of producing all the energy it needs, without having to depend on other nations around the world for its coal, oil, and natural gas. Various presidents have recommended different routes to that objective—through conservation, increased development of U.S. natural resources, or an expansion of alternative energy forms such as wind and solar power—but all have agreed on the desirability of energy independence.

Until recently, most experts in the field have expressed doubts as to whether the United States could ever achieve that goal or, if it could, whether it could become energy independent within the foreseeable future. The nation was such a voracious consumer of energy, and its energy resources appeared to be so limited, that a goal of energy independence seemed to be out of reach to most people who studied the subject (which is not to mention the number of voices who asked whether energy independence was even a *smart* national policy to pursue).

Sometime after the turn of the new century, however, the discussion of energy independence took a whole new turn. Evaluations of the amount of oil and gas, in particular, that could be extracted from U.S. territories suddenly began to rise. That does not mean that more oil and gas was being produced in Earth's surface under the United States, but it did reflect the fact that new technologies were being developed that were capable of removing oil and gas that had previously (1) not been known to exist or (2) had been thought to be too tightly held underground for commercial extraction with technology known at the time.

Two major technologies were responsible for this change in outlook: directional (also known as horizontal) drilling and hydraulic fracturing (commonly known as fracking). When used in combination with each other (as they often are), the two technologies have proved to be successful in squeezing oil and gas out of deposits that had previously been ignored or abandoned. After years of decreasing oil and gas production in the United States, conditions began to change after 2000 at a rate such that national and international energy agencies were predicting that the United States would be energy independent by 2020 or earlier.

As exciting as these predictions were, they tended to ignore or play down a number of social, political, economic, environmental, health, and other issues associated with the use of fracking and horizontal drilling by the oil and gas industry. For example, the former process makes use of dozens or even hundreds of different kinds of chemicals to increase the efficiency of the fracking process. Some of these chemicals are known to have deleterious effect on the health of plants and animals, including humans. No sooner were reports coming in of huge flows of oil and gas from newly drilled wells than were those reports followed by complaints from people living near those oil and gas fields about respiratory problems, increased risks for cancer, psychological problems, disruption of the natural environment, and other damage to the standard of living.

The responses to these complaints ranged across the board. Some individuals and companies, especially those involved in the exploration for and extraction of gas and oil, tended to downplay the seriousness of these charges, raising questions in some cases as to how often people were imagining the conditions they reported. Other individuals and agencies suggested that some level of health and environmental "inconvenience" might be necessary to gain the full benefit of the new energy technology that had become available to the world. Still other individuals and associations argued that the risks associated with those benefits were too great to expand the use of fracking and horizontal drilling. In the most extreme of cases, those individuals and associations have eventually been successful in getting governmental agencies from the national level (France and Bulgaria, for example) to the state level, and especially to the local level of towns, cities, villages, and counties to ban the use of fracking within their boundaries.

Given the very recent rise in fracking activities in the United States (and a very few other countries in the world), it is not yet clear how the world will reach some form of accommodation between the arguments for and against fracking, or how the new technology will continue to be used under regulatory control that will satisfy the majority of the citizens of a city, state, and nation. Already governmental units at all levels are struggling with a plethora of legislative and administrative remedies for resolving this issue, and the courts have already become active in trying to resolve disputes between those who favor and those who oppose fracking operations.

The purpose of this book is to provide the reader with a detailed background of the history and development of the oil and gas industry from the earliest days of human civilization to the present moment, with special attention on the history and development of hydraulic fracturing and directional drilling (Chapter 1). The book also attempts to lay out the primary concerns that have been expressed about the use of fracking in the exploration for and extraction of oil and gas, focusing on the arguments in support of and opposed to a wider use of the technology (Chapter 2). The second half of the book provides readers with a variety of resources designed to assist one in further research on the topic, such as an annotated bibliography of books, articles, and Internet resources on the topic, a chronology of important events in the history of fracking, a glossary of common terms used in the discussion of fracking and horizontal drilling, and profiles of important individuals and institutions active in the field of fracking. One chapter (Chapter 3) also provides selected individuals with an opportunity to express their views on some specific aspect of the fracking debate.

Fracking

> *The problem of peak oil remains. . . . In our opinion, it will be very difficult to raise the oil production above 95 million barrels/day*
>
> (Desmarest 2010)

This view on the worldwide availability of oil was offered by Thierry Desmarest, chairman of the board of the Total oil company at the World Economic Forum held at Davos, Switzerland, in January 2010. *Peak oil* is a concept that arose out of the research of Shell corporation geologist M. King Hubbert in the mid-1950s. In a series of scholarly papers, Hubbert suggested that fossil fuel production must inevitably follow a predictable pattern that later became known as the Hubbert curve. According to this theory, the production of coal, oil, and natural gas increases from essentially zero at some point in history to some maximum point, after which production decreases, until it again reaches a very low or near-zero level. The Hubbert curve has essentially the same shape as the well-known bell curve (also known as a normal curve or Gaussian function) that describes the behavior of many natural phenomena.

An oil well in Oklahoma, ca. 1922. (Library of Congress)

Hubbert's argument was simple in some regards. Scientists know that fossil fuels were produced over many millions of years early in Earth history as the result of a combination of geological processes. At some point in time, those processes ceased to exist, and the fossil fuels were no longer being produced. At a much more recent period in history, humans discovered the existence of fossil fuels and their potential value for driving a number of useful operations. They began to extract and make use of these fuels during a period in modern history that has become known as the age of fossil fuels. The age of fossil fuels dates from about 1800, soon after the start of the Industrial Revolution, to at least the present day. During this two-century period, humans have extracted very large amounts of coal, oil, and natural gas from beneath Earth's surface, perhaps as much as half of all the fuels present under the crust.

At some point, many scientists believe, the amount of coal, oil, and natural gas that can be extracted from the Earth will reach a maximum, a level known as *peak oil*, *peak coal*, and *peak natural gas*. Beyond that point, the available reserves of all three fossil fuels will continue to decrease, such that humans will be able to withdraw smaller and smaller amounts of the fuels from Earth each year.

The consequences of such a pattern are probably self-evident. For at least 200 years, humans have been building a civilization based on the ready availability of fossil fuels. We have transportation systems, heating systems, industrial systems, military systems, housing systems, and other systems that operate primarily because coal, oil, and natural gas can be burned to produce energy. In addition, these fossil fuels have also become the raw material from which a seemingly endless list of synthetic products can be made. As our supplies of fossil fuels begin to diminish, dramatic changes in the ways in which humans live their lives may be necessary.

The point at which peak oil, peak coal, and peak natural gas are likely to occur, according to those who believe in this argument, differs from fuel to fuel and from nation to nation. According to one recent study, peak oil may already have occurred some time ago in countries such as the United States (1970), Canada (1973), Iran (1974), Romania (1976), and Indonesia (1977) or more recently in countries such as Equatorial Guinea (2011), Algeria (2012), Azerbaijan (2013), India (2015), and Sudan (2015). The same studies place the date for peak oil worldwide as 2014 (Nashawi, Malallah, and Al-Bisharah 2010).

Estimating peak coal is considerably more difficult, at least partly because of the different types of coal being mined (primarily anthracite, bituminous, and lignite). In 2007, however, the Energy Watch Group attempted to make some estimates with regard to the year in which coal production would peak in a number of major coal-producing nations. The group said that peak anthracite coal had already occurred in the United States in 1990 and that total coal production would probably peak sometime between 2020 and 2030. It estimated that total coal production had peaked in Germany in about 1985, although anthracite had peaked nearly 30 years earlier. Peak production in China is very difficult to predict, but may occur as early as about 20 years into the future, an event that is likely to mark the date of world peak coal also (Coal: Resources and Production 2007).

As with oil and coal, experts differ widely on the possibility and/or timing of a peak natural gas scenario. Some data suggest that peak gas may have already occurred in some countries. Romania, for example, has Europe's longest history of exploration and extraction of natural gas. But the country apparently reached peak gas in 1976 when it produced 29.8 billion cubic meters of the product. Today, production has dropped to about a third of that amount (Viforeanu, Wells, and Hodny 2003). A similar trend has occurred in the United Kingdom, where gas production peaked at just less than 4 billion cubic feet in 2000 and has since dropped to about 1.3 billion cubic feet in 2012 (BP Statistical Review of World Energy 2013, 22).

The term *peak energy* is sometimes used to assess that year or period when humans will have extracted the maximum amount of fossil fuels from Earth's surface. At that point, we will still be mining coal and drilling for oil and natural gas, but the total amount of these fuels obtained will continue to decrease over time. Civilization will be faced with the question as to how it will continue to meet its energy needs with an ever-decreasing supply of fossil fuels. A number of experts have attempted to estimate the year in which peak energy will occur, with estimated peak energy often estimated as occurring sometime between 2020 and 2040 (see, for example, Clugston 2014, Cooke 2014, Mórrígan 2014).

The theory of peak oil, coal, natural gas, and energy, however, is by no means universally accepted. In fact, many experts in the field of energy resources reject the concept, offering a range of objections from modest adjustments to the predictions that have been made to very strong complaints that pro-peakers just don't know what they are talking about. For example, the comment with which this chapter begins was made by Desmarest in response to an earlier remark by Khalid al Falih, chairman of Saudi Arabia's Aramco oil company, the world's largest oil corporation. Earlier at the Davos meeting, al Falih had disparaged the notion of peak oil, stating simply that "we don't believe in peak oil." Oil reserves were sufficient, he said, for production to continue growing well into the foreseeable future (Neo 2014).

A very important development has occurred in the last two decades that plays a critical role in the discussion about peak energy: the use of hydraulic fracturing as a method for gaining access to oil and natural gas reserves stored in shale rock in many parts of the world. Hydraulic fracturing is a form of mining in which a mixture of water and sand, along with various other chemicals, is injected into rock, causing the rock to break open and release oil and natural gas stored within it. Hydraulic fracturing is also widely known by the term *fracking*. Fracking was, for all practical purposes, an irrelevant technology in the exploration and extraction of oil and gas prior to 2000. In that year, about 300,000 barrels of oil and about 2 billion cubic feet of natural gas per day were produced by the fracking process. Those numbers increased to about 3.25 million barrels of oil and 32 billion cubic feet of natural gas per day by 2013 (Sieminski 2014, 4). As of 2011, the latest date for which estimates are available, hydraulic fracturing accounted for about one-third of all the natural gas produced in the United States (SEAB Shale Gas Production Subcommittee Ninety-Day Report 2014, 6). Whether hydraulic fracturing can significantly affect the problem of peak energy in the United States and the rest of the world is a topic of considerable discussion today, and it is the main theme of this book.

The Genesis of Fossil Fuels

In order to better understand the issues posed by hydraulic fracturing, it is worthwhile to review some basic background information about the fossil fuels: how they were formed; when and how humans learned to extract them from the Earth; what challenges they pose for the economics, politics, social issues, and other aspects of civilization today; and how fracking relates to all these issues.

On one level, explaining the origin of coal, oil, and natural gas is a relatively simple and straightforward process. Before reviewing that process, however, we should point out that this "relatively simple and straightforward process" masks a far more complex set of physical, chemical, geological, and biological changes that petroleum engineers have only barely begun to understand. But let's start with the simpler explanation first.

Some types of organisms are capable of converting the energy of sunlight directly into organic material. For example, plants use this energy in the process of photosynthesis, by which carbon dioxide and water are converted into complex carbohydrates. A number of microorganisms are capable of carrying out the same types of chemical reactions. Organisms able to carry out these changes are called *primary producers*, or simply *producers*. (By contrast, organisms that are not able to manufacture their own foods, as do producers, are called *consumers*.)

When any organism dies, it eventually decomposes and, under most circumstances, the organic compounds of which it is made are converted back to the basic substances of which they were made, primarily carbon dioxide and water. This process of decay and decomposition follow this pathway, however, only in the presence of oxygen. In the absence of oxygen, a different type of decay and decomposition takes place, a process known as *anaerobic decomposition* (i.e., decomposition without oxygen). The products of the anaerobic decay of organic matter are quite different from those of aerobic (with oxygen) decay. They include primarily hydrocarbons, organic compounds made only of carbon and hydrogen, and of pure or nearly pure carbon. Notice that these decay products do not contain oxygen because no oxygen was available during the decay process.

Now imagine what the Earth would have looked like during its early history, say a few hundred million years ago. The planet was still relatively young at that point, and its geology and climate were very different from what they are today. Temperatures were higher, there was more moisture present, and a variety of simple life forms were abundant. The landscape was covered with swamps, bogs, lakes, and other watery environments.

When organisms died in such an environment, they might very well end up on dry land and decay as most organisms do today, with their bodies eventually decomposing into carbon dioxide and water. But there was also a high likelihood that they might die within a watery environment, sink to the bottom of that body of water, and decompose without access to oxygen. Scientists believe that high rates of anaerobic decay were common in this early Earth environment, resulting in a far greater likelihood that the decay of organisms would result in the formation of carbon and hydrocarbons than in the formation of carbon dioxide and water.

The second important factor contributing to the development of fossil fuel reserves was earth movements. Think of the range of earth movements that take place today: swamps eventually fill in and become meadows; landslides occur filling in lakes and ponds; earthquakes create cracks in the ground and/or move earth around. Events such as these in Earth's early history could easily have resulted in the burial of dead organic material under a few meters or more of ground. Over time, more and more layers of earth could have built up on top of all these dead bodies, increasing both pressure and temperature on them, accelerating the process by which their organic components were converted into carbon and hydrocarbons.

And that is essentially the story of how coal, oil, and natural gas were formed. In the case of coal, the dead organisms were usually trees and other plant material that died, fell into swamps, were covered with layers of earth, and gradually decomposed to form coal and coal-like products. The coal found in the Earth today varies from place to place depending on the conditions under which it was formed. If dead trees were exposed to very high pressures and temperatures, decomposition was almost complete; the trees' organic compounds were converted almost entirely to pure carbon. Coal formed by this process is called *anthracite* (or "hard coal"), a form of the fuel that consists of more than 90 percent pure carbon, along with very small amounts of hydrocarbons. *Bituminous* (or "soft coal") formed under less severe conditions, such that some of the original organic compounds have been only partially decomposed, to hydrocarbons, and others completely decomposed, to pure carbon. Finally, a form of coal called *lignite* consists of more hydrocarbons and less pure carbon than either anthracite or bituminous coal.

Scientists believe that the conditions under which dead plant material decays anaerobically as described in this scenario were most common during a geological period now known as the Carboniferous Period, which lasted from about 360 million years ago to about 300 million years ago. In fact, this geologic period got its name from the scenario outlined here, a period of "coal bearing," which is what the name means.

The process by which petroleum and natural gas were originally formed differs from that by which coal was formed in one regard because of the organisms involved. Instead of dead trees and plant material, the material from which oil and gas were formed was microorganisms. A number of experts today believe that the primary organisms responsible for the formation of oil and gas were the diatoms, a very large group of single-cell phytoplankton that live in virtually all types of watery environments.

Petroleum and natural gas formation begins when vast numbers of microorganisms in a lake or other body of water die and sink to the bottom. There they are covered with sand, clay, and other sediments, such that they decay without access to oxygen producing a mixture of liquid and gaseous hydrocarbons known as *kerogen*. As pressure and temperature increase in the microorganism-filled layer, the kerogen is gradually converted to petroleum (above 80°C) and natural gas (above 120°C).

While this conversion is under way, geologic changes are almost certainly occurring as well. The layer consisting of sand, clay, kerogen, and other materials is probably shifting and folding in response to earth movements. The surface of the layer probably consists of upward protrusions and downward depressions ("hills and valleys"). Thus, as oil and natural gas begin to form, they flow upward (because they are less dense) into the upward protrusions, forming pockets of liquid and gaseous fossil fuel. As the lake or other body of water dries out over millions of years, the oil and gas remain trapped within these *domes* that now consist of sandstone, shale, or some other rocky material. (For excellent animations of the oil- and gas-making process, see Oil Formation [two files].)

Many people have the image of an oil and gas reservoir as being similar to a "lake" of liquid oil lying within a dome, over which exists an "atmosphere" of natural gas. Such is not the case, however. Instead, the oil and gas are trapped in the pores of rock strata, such as sandstone or limestone. As an analogy, think of a sponge, the bottom part of which has been soaked with water, while the upper part remains dry. A petroleum reservoir is like that sponge in that the lower part of a rock layer is "soaked" with oil, trapped in countless numbers of tiny pores in the rock, while the upper part of the rock layer holds natural gas trapped in pores, just as air is trapped in the pores of a dry sponge. The challenge for oil companies (as discussed later in this chapter) is to find ways of drilling down into a rocky layer containing oil and gas and forcing those materials upward to the surface of the Earth.

Another contrast between the formation of coal and the formation of petroleum and natural gas is that, as noted above, the former appears to have occurred during one rather limited period in Earth history, the Carboniferous, which lasted only about 60 million years. By contrast, the formation of oil and gas appears to have continued over a long period of time, throughout a period now known to scientists as the Phanerozoic Eon. The term *phanerozoic* comes from the Greek expression meaning "visible life." It refers to the range of Earth history during which multicellular organisms first appeared and began to evolve. It began about 542 MYA (million years ago) and includes the Paleozoic, Mesozoic, and Cenozoic Eras.

Scientists now believe that oil and gas were formed sporadically throughout this period as a result of differing geological, climatic, and other conditions. For example, the first burst of oil and gas formation appears to have taken place during the Silurian Period, between about 440 and 400 MYA. Rocks containing kerogen and natural gas dating to this period can be found today in the Middle East, North Africa, and parts of the United States. Researchers have estimated that about 9 percent of all the oil and gas currently available on Earth was produced during this period (A Crude Story 2014; Sorkhabi 2014).

The most productive periods for oil and gas appear to have been the Jurassic (about 200 to 140 MYA) and Cretaceous

(140 to 65 MYA) Periods, during which about 25 percent and 29 percent, respectively, of all oil and gas were produced. (These periods are perhaps best known as the age of dinosaurs, which explains the association that is sometimes made with the death of these reptiles with the formation of fossil fuels when, in fact, no such connection actually exists.) The Jurassic and Cretaceous Periods were also characterized by some dramatic geological and climatic changes. For example, the period was one during which the planet's original protocontinent, Pangaea, began to break up and drift apart, creating the continents that we know today. Thus, dramatic earth movements that would contribute to the formation and isolation of oil and gas deposits were probably fairly common. At the same time, the Earth's climate began to warm, producing a hothouse atmosphere in which plant growth thrived and bodies of water expanded to play host to unimaginably vast numbers of protozoa from which oil and gas were eventually to develop. Source rocks from these periods, primarily shale and limestone, are found today in widespread regions of the Earth, including the Middle East, the North Sea, the Gulf of Mexico, and western Siberia. The rocks are rich in kerogen, petroleum, and natural gas (A Crude Story 2014; Sorkhabi 2014).

The most recent periods during which oil and gas were formed were the Miocene (about 5 to 23 MYA) and Oligocene (23 to 34 MYA) Epochs. Because these events are so recent in geologic terms, oil and gas formed during these periods is sometimes called *new oil and gas*. (For a summary of oil and gas formed in various geologic periods, see Emery and Myers 1996, Figure 11.2, page 240.)

The most familiar type of reservoir is the dome-shaped structure that contains liquid petroleum above which lies an atmosphere of natural gas, but oil and natural gas that occurs within the Earth today is present in a variety of forms, depending on the method by which it was produced and changes through which it has gone.

Oil and Gas Resources

We will now briefly discuss terminology. The terms *crude oil*, *petroleum*, and *oil* are often used interchangeably in discussions about oil and gas. The terms are actually somewhat different. *Crude oil* refers to the liquid mixture of hydrocarbons that occurs in reservoirs beneath the Earth's surface, the liquid that has been mentioned earlier in this book as *oil*. The two terms describe essentially the same material. Once crude oil is removed from the Earth and treated so as to make it useful for a variety of applications, those products are then known as *petroleum products*. Finally, the word *petroleum* itself refers to crude oil plus the petroleum products made from it. In practical situations, the words *oil* and *petroleum* are often used interchangeably, although such a practice is not strictly correct (Glossary 2014).

The dome-shaped trap described above is more technically described as a *structural trap*, any type of geological structure that allows oil and/or natural gas to collect within a restricted area. (For an explanation of structural traps, see Petroleum Traps 2014.) Structural traps may contain liquid hydrocarbons (oil) only, gaseous hydrocarbons (natural gas) only, or some combination of the two forms. When the oil and gas are mixed with each other in solution, the natural gas is often called *dissolved natural gas*. When the two are separate from each other, the gas is said to be *associated natural gas*.

Oil and natural gas reserves do not occur uniformly across the planet's surface. They are present more abundantly—often much more abundantly—in some places than in others. Most people know, for example, that the Middle East contains particularly rich oil and gas reserves, as do certain parts of the United States, Mexico, Venezuela, and offshore regions such as the North Sea and Gulf of Mexico. Any region that has extensive oil and/or gas reserves spread over an extended area is called an *oil field*. According to some authorities, there may be as many as almost a thousand very large oil fields on the planet (where an oil field is defined in this case as a region with at least 500 million barrels of recoverable oil).

The largest oil field in the world is said to be the Ghawar field in Saudi Arabia. The field first started producing oil and gas in 1951 and is said by some to have reached peak production in 2005, although that view has been disputed by some experts (see, for example, Tech Talk—Current Oil Production and the Future of Ghawar 2014). Currently, the field yields about 5 million barrels of oil and 2 billion cubic feet of natural gas per day, accounting for about 7 percent of the world's output of oil and gas (Popescu 2014). According to some observers, there may be as many as 3,000 distinct oil wells in the Ghawar field (How Many Wells in Ghawar? 2014). Other very large oil and gas fields around the world are the Burgan field in Kuwait (1.7 million barrels of oil and 550 million cubic feet of gas per day), Ferdows, Mound, and Zagheh fields in Iran (output not known), Sugar Loaf field in Brazil (under development), and Cantarell field in Mexico (500,000 barrels of oil; natural gas not available). (The 10 Largest Oil Deposits in the World 2014). The top 25 oil and gas fields in the United States, based on proved reserves, are shown in Tables 1.1 and 1.2.

Geologists and specialists in related fields also refer commonly to another concept related to fossil fuel reserves, the *play*. Although that term has been in use for a long time, it still has no widely accepted and clearly defined meaning. One possible definition that has been proposed is the following:

> A play is defined as a set of known or postulated oil and or gas accumulations sharing similar geologic, geographic, and temporal properties, such as source rock, migration pathways, timing, trapping mechanism, and hydrocarbon type. (Biewick, Gunther, and Skinner 2002)

Plays are often named to reflect the geologic period during which their oil and gas were formed, for example, a Devonian play, Silurian play, or Mississippian play.

Table 1.1 Top 25 U.S. Crude Oil Fields as Ranked by Estimated 2009 Proved Reserves

Rank	Field	State	Proved Reserves[1]	Year Discovered
1	Prudhoe Bay	Alaska	95.9	1967
2	Spraberry Trend	Texas	39.6	1949
3	Mississippi Canyon Block 807	FG[2]	66.0	1989
4	Mississippi Canyon Block 778	FG	73.7	1999
5	Belridge South	California	32.6	1911
6	Kuparuk River	Alaska	38.0	1969
7	Wasson	Texas	22.1	1937
8	Green Canyon Block 743	FG	38.3	1998
9	Midway-Sunset	California	34.1	1901
10	Elk Hills	California	13.4	1919
11	Kern River	California	28.8	1899
12	Green Canyon Block 826	FG	23.6	1998
13	Wattenberg	Colorado	15.6	1970
14	Slaughter	Texas	10.3	1937
15	Cedar Hills	Montana/ North Dakota/ South Dakota	20.8	1951
16	Green Canyon Block 640	FG	22.6	2002
17	Milne Point	Alaska	10.6	1982
18	Salt Creek	Wyoming	3.7	1889
19	Wilmington	California	13.6	1932
20	Levelland	Texas	7.6	1945
21	Seminole	Texas	6.9	1936
22	Monument Butte	Utah	4.6	1964
23	Cymric	California	17.6	1916
24	Goldsmith	Texas	6.0	1935
25	Orion	Alaska	0.0	2002

[1]Million barrels.

[2]Federal offshore, Gulf of Mexico.

FG = Federal offshore source, Gulf of Mexico

Source: Top 100 Oil and Gas Fields of 2009, U.S. Energy Information Administration, Table 1, http://www.eia.gov/pub/oil_gas/natural_gas/data_publications/crude_oil_natural_gas_reserves/current/pdf/top100fields.pdf. Accessed on June 7, 2014.

Table 1.2 Top 25 U.S. Natural Gas Fields Ranked by Estimated 2009 Proved Reserves

Rank	Field	State	Proved Reserves[1]	Year Discovered
1	Newark East	Texas	1794.6	1981
2	San Juan Basin Gas Area	Colorado/New Mexico	1295.2	1927
3	Pinedale	Wyoming	487.8	1955
4	Haynesville Shale Unit	Louisiana	203.7	2008
5	B-43 (Fayetteville)	Arkansas	516.8	2004
6	Prudhoe Bay	Alaska	166.7	1967
7	Jonah	Wyoming	391.1	1977
8	Natural Buttes	Utah	224.4	1940
9	Hugoton Gas Area	Kansas/Oklahoma/ Texas	328.1	1922
10	Wattenberg	Colorado	195.3	1970
11	Spraberry Trend Area	Texas	113.8	1949
12	Carthage	Texas	273.0	1936
13	Mamm Creek	Colorado	111.4	1959
14	Fogarty Creek	Wyoming	34.7	1975
15	Antrim	Michigan	125.6	1965
16	Grand Valley	Colorado	191.0	1985
17	PRB Coal Bed	Montana/Wyoming	558.9	1915
18	Big Sandy	Kentucky/Wets Virginia	70.6	1926
19	Lake Ridge	Wyoming	15.7	1981
20	Oakwood	Virginia	136.7	1990
21	Raton Basin Gas Area	Colorado/New Mexico	140.4	1998
22	Lower Mobile Bay Area	Alabama/FG[2]	156.1	1979
23	Madden	Wyoming	88.6	1968
24	Rulison	Colorado	127.2	1958
25	Carthage North	Texas	24.9	1966

[1]Billion cubic feet.

[2]Federal offshore, Gulf of Mexico.

FG = Federal offshore source, Gulf of Mexico

Source: Top 100 Oil and Gas Fields of 2009, U. S. Energy Information Administration, Table 2, http://www.eia.gov/pub/oil_gas/natural_gas/data_publications/crude_oil_natural_gas_reserves/current/pdf/top100fields.pdf. Accessed on June 7, 2014.

One of the most commonly mentioned oil and gas plays today is the Marcellus play that lies beneath about 34 million acres of land in the mostly northeastern United States within the states of New York, Pennsylvania, Ohio, Kentucky, Tennessee, Virginia, and West Virginia, and, to a small extent, Maryland, Alabama, and Georgia, as well as a portion of Lake Erie. (For a map of the Marcellus play, see http://www.shalereporter.com/maps/pdf_911fdc84-67e4-11e1-8dab-0019bb30f31a.html.) The Marcellus play is believed to contain 500 trillion cubic feet of natural gas, of which about 10 percent, or 50 trillion cubic feet, is thought to be economically recoverable (Engelder and Lash 2014). Although there is no accurate count of the number of wells that have been sunk into the Marcellus play at this point in time, there are strong indications that it has been and will continue to be the locus of high activity among oil and gas exploration companies. For example, the number of wells drilled into the Marcellus play in Pennsylvania alone rose from 27 in 2007 to 2,073 in 2011, only four years later (Marcellus Shale–Appalachian Basin Natural Gas Play 2014).

Especially with the increased availability of hydraulic fracturing as a way of collecting shale oil and gas, interest in fossil fuel plays around the world has greatly increased in many countries. Today, nations such as Mexico, Argentina, Canada, India, China, and Russia have all discovered very large gas and oil plays, which hold the promise as very significant sources of these materials in the future. For maps that show the size and location of some of these plays, see http://pacwestcp.com/education/regional-maps/.

As an example of some of the extraordinary plays that have been discovered, consider the Bazhenov play in western Siberia. The play runs from the Arctic Sea in the north to the borders of Kazakhstan in the south, covering an area of about 1 million square kilometers (about 400,000 square miles). The U.S. Geological Survey estimates that the play may contain up to 1,243 billion barrels of oil, of which 74.6 billion barrels are economically recoverable, and 1,920 trillion cubic feet of gas, of which 285 trillion cubic feet is economically recoverable (Technically Recoverable Shale Oil and Shale Gas Resources 2013, IX 2). Note that these numbers are four to five times the estimates for the Marcellus shale.

Terminology

Petroleum experts and companies usually like to describe the amount of oil and/or gas that can be recovered from a well, field, or play. They use terms such as *technically recoverable*, *discovered unrecoverable*, and *risked* to describe resources. Such terms reflect one basic fact about oil and gas reserves: It is often difficult not only to know how much of the product is present in a resource, but also how much of that product can actually be extracted from the Earth. It is not uncommon to discover a new oil and/or gas reserve, which is suspected to contain very large quantities of the product. But for any number of reasons, engineers are fairly certain that only a relatively small fraction of that product can actually be extracted from the resource.

The problem of petroleum terminology is made more complex by the fact that many different agencies use such terms, and those agencies frequently do not agree with each other. One article on this issue lists just a few of the agencies for whom the prediction of oil and gas reserves is essential and who, therefore, have their own systems for describing the size of those reserves: the U.S. Geological Survey, American Association of Petroleum Geologists, Society of Petroleum Engineers, Society of Petroleum Evaluation Engineers, World Petroleum Congress, Securities and Exchange Commission, Internal Revenue Service, and the United Nations Framework Classification (Oil and Natural Gas 'Reserves'—Definitions Matter 2014). The following list provides definitions for at least some of the terms used in discussing the size of reserves in an oil or gas well, field, or play. But remember that for each term, alternative definitions are also available.

Proved (or proven) reserves: The estimated quantities of oil and/or gas that experts believe with reasonable certainty to be recoverable in future years from known reservoirs under current economic and operating conditions.

Probable reserves: The estimated amount of oil and/or gas that experts believe can be recovered from a resource with something less than high probability. In some cases, the cutoff point is taken as a 50 percent probability, in which case the reserves are called P50 reserves.

Possible (or inferred) reserves: The estimated amount of oil and/or gas that experts believe can be recovered from a resource with a relatively low level of probability, often around 10 to 20 percent likelihood, given current economic and operating conditions.

Ultimately recoverable resource: An estimate of the total amount of oil and/or gas that will ever be extracted from the Earth.

Technically recoverable resource: An estimate of the amount of oil and/or gas that can be recovered using the best available current technology *without concern* as to the economic viability of such efforts.

Undiscovered resources: An estimate of the amount of oil and/or gas that experts *guess* is likely to be available in a region where the product has not yet actually been found. A subdivision of this term is *technically recoverable undiscovered resource*, which refers to the amount of that as-yet undiscovered resource is likely to be recoverable using current technology.

Contingent resources: Estimates of the amount of oil and/or gas that can be recovered from a resource that is not yet being used for any one of a number of economic, legal, regulatory, environmental, or other reasons.

Prospective resources: An estimate of oil and gas resources that have not yet been discovered, taking into consideration new forms of technology that may make the extraction of those resources technically and economically possible.

Risked resources: Oil and/or gas reserves are reserves that are possible or probable, but that have not yet been produced.

Unrisked resources: Oil and/or gas reserves are reserves that have already been developed. (An important standard source of most definitions relating to oil and gas reserves is Petroleum Resources Management System 2007.)

Exploring for Oil and Gas

Most oil and gas reserves today are buried deep beneath Earth's surface. As of late 2014, the deepest oil well in the world is the Chayvo well, Z-44, off the east coast of Russia's Sakhalin Island with a depth of 12,376 meters 40,604 feet. But historically, oil and gas reserves have been found at virtually any depth, including just beneath Earth's surface and even, in some cases, on top of the ground. A one-time common experience in the search for oil was immortalized in a 1960s television comedy, *The Beverly Hillbillies*, in which a family struck oil simply by driving a post into the ground and "up came the bubblin' crude." Such easy discoveries of shallow reservoirs of oil and gas are, however, long gone, and scientists use a variety of sophisticated techniques to find reserves buried many thousands of meters underground.

The most common method for searching for oil and gas reserves today is seismic surveying. In a seismic survey, a shock wave is created at ground level usually by means of setting off a small explosion or by vibrating heavy plates carried by large trucks. The waves travel below the ground and are reflected back to the surface, where they are recorded by means of specialized devices similar to microphones called *geophones*. The data recorded by the geophones are then analyzed to obtain an image of rock structures underground. These images allow a geologist to find regions where oil and/or gas are likely to be found. For example, fractures in underground structures often provide conduits through which oil and gas are able to flow and collect with porous layers of rocks. A seismic image often indicates the presence of such a fracture. Today, three-dimensional seismic images can be produced that provide even more detailed and precise images of underground structures and, hence, more reliable information about the possible location of petroleum reserves.

Seismic imaging can also be conducted on the water. Ships drag devices that release bursts of compressed air into the water. The shock waves produced by the compressed air are transmitted through the rocky structure under the water and are reflected back to the water's surface, where they are captured by *hydrophones*, water versions of a geophone. The hydrophone images can then be analyzed in essentially the same way as with land-based seismic systems. (For an animated explanation of seismic imaging, see http://www.youtube.com/watch?v=hxJa7EvYoFI or http://www.youtube.com/watch?v=22m27MhzSQs.)

Another technology available for oil and gas exploration takes advantage of the Earth's natural gravitational and magnetic fields. As everyone knows, gravity is a force that attracts objects to Earth's center. It is a force that is relatively constant at every point on Earth's surface. However, very small variations in the gravitational force exist at different points on Earth's surface for a variety of reasons, including differing underground rock structures and variations in the chemical composition of rocks. Geologists make use of these facts to look for regions where oil and gas reservoirs are likely to be found. They use highly sensitive measuring devices known as *gravitometers* (or *gravimeters*) to measure the precise value of the gravitational attraction at various points in a region being explored. The data received from these measurements are then fed into a computer system, where programs are available for the conversion of the data to maps of the underground regions. These maps provide geologists with a clue as to whether or not the region is worth exploring in more detail—such as sinking a trial hole—in the search for petroleum.

Just so that no one is offended with the terminology used here:

A *geologist* is someone who studies the structure of the Earth, its history, and physical and chemical changes that take place within the Earth.

A *geophysicist* focuses on the physical characteristics of the Earth, such as its gravitational and magnetic fields, plate tectonics, earth movements, and the structure of the mantle and core.

A *petrologist* is a geologist who specializes in the study of the composition, origin, distribution, and structure of rocks.

A *petroleum geologist* is someone who studies the origin, occurrence, exploration, and extraction of liquid and gaseous hydrocarbons.

(For more information on this topic, see http://www.radford.edu/content/csat/home/geology/faq.html.)

The principles that underlie the use of gravitational forces in oil and gas exploration are also used in magnetic surveys. Much of what has been said about gravitational surveys also applies to magnetic surveys because Earth's magnetic field varies in very subtle ways from place to place in the same way its gravitational field varies. (For animations of gravitational and magnetic imaging, see http://www.youtube.com/watch?v=9P6GEpxFtSY and http://www.youtube.com/watch?v=AZyNIGFHsE4.)

Yet another technology used in oil and gas exploration is called electromagnetic (EM) surveying. This technology takes advantage of the fact that various types of rock conduct an electrical current more or less easily. In particular, porous rocks that are most likely to hold crude oil and natural gas tend to be relatively poor conductors of electricity compared to other types of rock. EM surveying involves sending an electrical current into underground strata and then measuring the resistivity of the rocks through which that current flows. Resistivity is the reverse of conductivity and is easier to measure than the flow of electricity. Detecting devices are able to distinguish very small changes in the resistivity of rock and, hence, act as predictors of places where drilling is or is not likely to be successful. EM surveying is generally used in conjunction with and to confirm the results of seismic surveys.

A final tool used in the exploration for oil and gas is remote sensing. The term *remote sensing* refers to the collection of data about objects on Earth's surface using aircrafts or satellites traveling high above the Earth. Remote sensing is often used as a preliminary technology to locate areas where the probability of finding oil and/or gas is good based on rock types and formation, fractures in the Earth's surface, microseepage of oil at Earth's surface, and other surface features. (For a good review of the use of remote sensing in petroleum exploration, see Finding Oil and Gas from Space 2014.)

Drilling Technology

Finding reservoirs of crude oil and natural gas is, of course, only the first step in a long process of putting gasoline into the tank of a car or feeding natural gas into a home heating system. The next step is finding a way to remove oil and gas from the ground. In the most general sense, that process generally involves drilling into the ground through layers of rock until one reaches a layer in which crude oil and/or natural gas is trapped. When that layer is reached, the oil or gas finally has a conduit—after many millions of years—by which it can escape to the Earth's upper surface. The oil and/or gas then begins to flow upward, driven by the fluid and gas pressure that has accumulated in the rock layer since its formation. The amount of that pressure available determines how easily the oil and/or gas escapes from the well. In many cases, the pressure is so great that the oil and/or gas come shooting out of the drill hole in the form of a *gusher* an explosion of gas and liquid that can shoot many meters into the air. For well drillers, the appearance of a gusher during a drilling operation can be one of the most exciting experiences of a lifetime.

Preparation for drilling begins weeks or months before work can actually start in the field with the process of purchasing or leasing the land to be used, obtaining all necessary documents, clearing the land where drilling is to occur, arranging for living and working conditions for the drill crew, taking such actions as may be necessary to protect the surrounding environment from damage by the drilling process, and making other such preparatory arrangements. All of these activities plus the drilling process itself requires a variety of specialized companies, each with its own specific task in the preparation and drilling process.

When the site has been prepared the drilling rig is brought into position for drilling. Drilling rigs differ widely in height and other specifications in order to provide the framework needed for any particular drilling job. Actual drilling begins with the creation of a *conductor hole* or *starter hole*, which holds the upper portion of the drilling system. The conductor hole has a large diameter of up to 20 inches or more and is lined with pipe. It may extend a few dozen or a few hundred feet into the ground, although some systems require no conductor system at all. When drilling has reached the bottom of the conductor hole, a second hole of smaller diameter is begun, the *surface hole*. Like the conductor hole, the surface hole is line with metal pipe to provide the sides of the hole from collapsing in on itself. The surface hole may extend to a depth of as much as 3,500 feet. Beneath the surface hole, yet a third section is added, the *intermediate hole*, which extends to a depth of nearly 10,000 feet, after which the final section, the *production hole* is drilled. Depending on the actual depth of the petroleum reservoir, the presence and/or dimensions of these sections may differ from hole to hole.

The "business" part of the drilling system is the *drill string*, which consists of three major parts: the bottom hole assembly (BHA), transition pipes, and drill pipes. The bottom hole assembly contains the drill bit, collars that keep parts in place, and other components, such as those needed to keep mud flowing through the system. The drill pipes are the pipes that run from the top of the well downward into the well until they reach the transition pipes, through which they are connected to the BHA. Drill pipes are thick steel pipes about 10 meters in length that are added to the well as the drill digs deeper into the earth. The whole piping system is held in place by pumping cement into the space between the pipes and the surrounding rock, where it hardens and forms a permanent grip on the piping system. Drilling continues until evidence of oil or gas begins to appear in the effluent from the drilling process. At this point, the lowest portions of the drill string are removed and a perforated pipe is attached. The holes in the pipe provide an outlet through which oil and gas can escape upwards through the pipes. At that point, a control mechanism called a *Christmas tree* is attached at the top of the pipe to direct the flow of oil and/or gas into storage tanks.

When the reservoir is reached, oil and/or gas begin to flow upward to the top of the piping system where they are captured. Historically, the oil produced in a well was by far more important commercially than was the gas. In fact, the natural gas obtained from a well was often simply just released to the atmosphere, or, more commonly, burned in a process known as *flaring*. Flaring allowed a company to get rid of a waste product in which it was not interested at little or no cost and with relatively little technical and environmental problems.

Of course, flaring is an enormously waste of what oil companies now realize is a very valuable product, but the arguments for flaring have traditionally been—and to some extent still are—convincing. One reason for flaring has been that the gas pressure that builds up in the drilling system can pose a safety threat to the equipment and crew working at the rig. Another reason is that the amount of natural gas produced by a well may be too small to justify the construction of a separate system for collecting and storing the gas. Over the years, oil companies have tended to reduce the amount of gas flaring at their wells, choosing instead to capture, store, use, and sell the captured natural gas. But flaring is still widely popular in many locations, with the main justification usually being the high cost of building systems for dealing with the natural gas. That is, it is simply cheaper to burn off the natural gas than it is to build a system for capturing and storing the product (Curtis and Ware 2014). As a consequence, recent studies have shown that oil companies are still venting and flaring very large amounts of natural gas—140 billion cubic meters in 2011—annually around the world. The loss of natural gas by venting and flaring in 2011, for example, was sufficient to supply all the natural gas used by all of Central and South America in one year (World Bank Sees Warning Sign in Gas Flaring Increase 2012).

One of the key elements in any drilling operation is the *drilling fluid*, also known as the *drilling mud*. Drilling fluid is a complex aqueous (water-based) mixture consisting primarily of bentonite clay, an aluminum silicate compound with variable composition, along with other possible components, such as calcium carbonate (chalk), barium sulfate (barite), iron oxide (hematite), various types of gums, starch, carboxymethylcellulose, and other substances. The precise composition of drilling fluid is determined by the specific needs of each drilling operation.

Drilling fluid is injected into the well through the middle of the drill string. It passes out of the bottom of the drill string, around the rotating drill bit, and then back out of the system through the space between the inner and outer pipes of the drill string. In its transport through the drilling system, the drilling fluid plays a number of roles, the most important of which is the removal of cuttings produced at the bottom of the hole by the drill bit. These cuttings become suspended in the drilling fluid and are carried to the top of the borehole, where they can be removed. The drilling fluid also acts as a heat exchange fluid, keeping the drill bit and other moving parts of the system cool during the drilling operation. The drilling fluid also acts as a pressure regulator within the drill pipes, ensuring that changing pressures on the pipes does not cause their collapse. (A diagram showing the operation of the drilling fluid system is available at http://www.kgs.ku.edu/Publications/Oil/primer12.html.)

The drilling process is a very complicated system, with far more elements than can be included in this section. For more detailed descriptions and animations of the drilling process see http://www.metacafe.com/watch/yt-cHfA4P8MMYc/drilling_for_oil_in_illinois/,http://www.wikihow.com/Drill-an-Oil-Well, and http://science.howstuffworks.com/environmental/energy/oil-drilling3.htm through http://science.howstuffworks.com/environmental/energy/oil-drilling5.htm.

Petroleum Chemistry

The material obtained from an oil well consists almost entirely of hydrocarbons, chemical compounds that contain only carbon and hydrogen. Methane is the simplest hydrocarbon. Its chemical formula, CH_4 indicates that a molecule of methane consists of one carbon atom, to which four hydrogen atoms are attached. Methane is by far the most common compound found in natural gas, making up about 95 percent of that product. The next heavier hydrocarbons are ethane (C_2H_6), propane (C_3H_8), and butane (C_4H_{10}). Butane is an example of a hydrocarbon that can exist in two forms, or *isomers*. The two isomers of butane contain four carbon atoms and ten hydrogen atoms, but the carbon atoms are arranged in a different pattern in the two isomers. The more carbon atoms present in a hydrocarbon, the more isomers are possible. The hydrocarbon called eicosane ($C_{20}H_{42}$), for example, had 366,319 isomers. The possibility of isomers helps explain why the chemical composition of oil and gas is so complex.

Another reason for the chemical complexity of petroleum is that carbon atoms in a hydrocarbon can be arranged in a variety of ways. Compounds in which the carbon atoms are all arranged in a straight line, in a branched arrangement, or in a ring are called *aliphatic hydrocarbons.* Aliphatic hydrocarbons in which the carbon atoms form a ring are also called *alicyclic hydrocarbons*. Cyclobutane, for example, consists of four carbon atoms arranged in a ring, with two hydrogen atoms attached to each carbon atom. Aliphatic hydrocarbons may also be *unsaturated*—that is, containing fewer hydrogen atoms than the maximum possible. For example, the compound known as ethylene (or ethene) has a chemical formula of C_2H_4. Comparing this compound to ethane (C_2H_6), it is obvious that ethylene has two fewer hydrogens than the maximum number possible with two carbon atoms. Therefore, ethylene is said to be unsaturated. Unsaturated compounds like ethylene are also called *alkenes* compared to saturated compounds like ethane, which is an *alkane*.

Another type of hydrocarbon is called an *aromatic hydrocarbon*. Aromatic hydrocarbons are alike in that they all contain a very characteristic structure called a benzene ring. Some familiar aromatic hydrocarbons are benzene, toluene, xylene, biphenyl, phenol, aniline, naphthalene, and anthracene.

The fact that hydrocarbons can exist in so many different forms explains the fact that crude oil contains so many—thousands of—different compounds. Faced with this reality, petroleum chemists spent relatively little time focusing on specific compounds present in crude oil, such as benzene or ethylene, and instead concentrate on segments, or fractions, of crude oil that consist of a few hundred different compounds with similar properties. The process of separating crude oil into a half dozen or so major segments is called *fractionating*, a process that takes place in tall towers called *fractionating towers* or *fractionating columns*. A fractionating tower is a cylindrical metal tower that may range anywhere from 6 to about 60 meters in height. It is divided into about a half dozen "floors" and contains an outlet pipe at each floor. The tower is heated at the bottom so that each floor in the tower is at a different temperature. Floors near the bottom are very hot because they are close to the source of heat, while floors near the top are much cooler.

When crude oil is added to the tower, it sinks to the bottom where it is heated. Almost all of the components of the crude oil are vaporized by the heat and begin to rise upward in the tower. Substances with very high boiling points may rise no more than one floor, while those with lower boiling points will rise until they are cool enough to change back to a liquid. Components with the lowest boiling points, then, may rise to the highest floor before they condense and change back to a liquid.

Some constituents of crude oil are dissolved gases (dissolved natural gas) that evaporate when the crude oil is heated in the fractionating tower. These so-called refinery gases rise to the top of the tower and are vented out through a pipe in the top of the tower. Another group of compounds, called the *gasoline* or *petrol* fraction, have a boiling point of somewhere around 70°C. They are vented from the tower at the level just below the top. A third group of compounds, called the *naphtha* fraction, condenses at a boiling point of about 150°C. The next lower levels capture fractions called *kerosene* and/or *jet fuel* (about 200°C), *diesel* and *fuel oil* (about 250°C), and *lubricating oil* (about 300°C). The lowest level of the tower holds the fraction with the highest boiling point, usually more than 400°C, and consists of asphalt, bitumen, tar, waxes, and other heavy liquids and semisolids. (Note that the classification and naming of fractions in a tower differs from location to location, company to company, and tower to tower. For an animation of a fractionating tower, see http://www.youtube.com/watch?v=GEJyiVq2NMM&index=11&list=PLOzrQUZ99ZqsarkxU_g6aafHLSkxmTkgb.)

As the preceding discussion suggests, crude oil can be divided into a number of fractions that have a wide variety of uses, from paving of highways (asphalt) to lubrication of machinery (lubricating oil) to coatings and coverings (wax). But the most important chemical property of the vast majority of petroleum fractions and compounds is combustion. The refinery gas, gasoline, jet fuel, and diesel fuel fractions are all burned in automobiles, trucks, jet airplanes, home heating systems, industrial plants, and many other locations throughout society. It is these applications of petroleum that constitute the reason that this substance is so enormously valuable to modern society. Nearly half (47 percent) of all crude oil is converted in one way or another to make gasoline, with another quarter (23 percent) used for diesel fuel and heating oil, and 10 percent for jet fuel (Oil: Crude and Petroleum Products 2014). So combustion is by far the most important end product of crude oil.

An additional use of crude oil, albeit a very small one, is the production of petrochemicals. The term *petrochemical* applies to any substance that can be made from one of the crude oil fractions. The list of petrochemicals produced today is almost limitless and includes items such as aspirin, candles, compact discs, crayons, fertilizers, footballs, garbage bags, heart valve replacements, perfumes, photographic films, safety glass, toothpaste, and vitamin capsules. Less than 5 percent of all the crude oil processed in the United States goes to the production of petrochemicals. But the number of petrochemical products and their ubiquitous role in society means that crude oil fractions have greater value than simply powering modern society.

A Brief History of Petroleum Exploration

As suggested by the preceding sections, drilling for oil and natural gas today is a highly complex, very sophisticated, and extremely expensive operation. Such has not always been the case, however. Petroleum reservoirs could once be found relatively close to the surface of the ground—or even on top of the ground—so it is not surprising that humans have known about and had access to crude oil and natural gas since the beginning of civilization.

Evidence for the earliest use of petroleum products dates to more than 40,000 years ago. Weapons and tools found at Neanderthal sites in modern-day Syria consisted of stone heads attached to shafts with bitumen, a viscous, dark, oily liquid with an odor similar to that of a freshly paved oil-covered road. The material was almost certainly obtained from surface pools of petroleum or *seeps*, places where oil and natural gas flow from beneath the ground through cracks to the surface. Probably the best known example of a petroleum seep today is the La Brea Tar Pits in Los Angeles, California. The material found in a seep is called by a variety of names in addition to bitumen, such as pitch, asphalt, and slime.

Probably the first precise description we have of bitumen and its natural sources comes from the writings of the Greek historian Herodotus in about 450 BCE who described the use of bitumen by the Babylonians as early as the second millennium BCE. He wrote that bitumen was used for a host of purposes, such as a mortar in the construction of brick walls and sailing ships, and for the waterproofing of baths and drainage systems (Ancient History Sourcebook 2014; Asphalt 2014). Other applications of bitumen included its use as a medicine (for treating skin disease and wounds and as a laxative), as a preservative in the preparation of mummies, and as a lining for baskets used to gather crops. In its natural or purified form, bitumen was also burned for use in illumination or as a weapon of war (Connan and Deschesne 1992).

The first real oil wells of which we know were those used by the Chinese in the fourth century CE. The date 347 CE is often given as the date on which the first such machine was built, although some question remains as to the accuracy of that year. The Chinese drill consisted of a system of bamboo pools with a primitive rock drilling bit that was pounded into the ground by the rocking motion of one of the bamboo poles. Such a device was apparently adequate for drilling to a depth of about 600 meters. (Kopey 2007; see an illustration of the machine in Figure 3 of this article.) The Chinese continued their development of drilling machines to the point that they required a somewhat complex system of bamboo pipes to move the oil recovered to salt water springs, where it was burned to evaporate salt water and recover solid salt (Oil and Gas: History 2014).

Possibly the most important region in the world in the history of the early development of petroleum exploration was the Baku region in Azerbaijan. Baku is situated on the Absheron Peninsula that projects eastward into the Caspian Sea. As early as the seventh century residents of the area were extracting oil from seepages that occurred throughout the area. These seepages were apparently so common that they often caught fire spontaneously producing huge "pillars of fire" that burned night and day. Historians suggest that the Zoroastrian religion, based on a worship of fire, had its beginnings in the region because of this startling natural phenomenon (Aliyev 2014; Oil and Gas: History 2014). Travelers who passed through the Absheron region over succeeding centuries often wrote about the huge production of oil and petroleum products and the uses to which they were put. For example, Marco Polo described in his writings how he saw oil being extracted on his visit to Baku in 1273 and his observation of its being used as a medicine and for illumination (Polo 1936). Other explorers who visited and wrote about the rich oil treasures of the Baku region included the Englishmen Thomas Bannister and Jeffrey Duckett in 1568, the Italian traveler Pietro della Valle in 1618, the Turkish visitor Evliya Çelebi in 1647, and the Dutch sailor and explorer Jan Struys in 1666 (History of the World Petroleum Industry 2014).

It seems almost inevitable, then, that the world's first modern drilling operations began in the Baku region in the mid-nineteenth century. Those operations were preceded in 1803 when a pair of wells was drilled by hand about 30 meters offshore in the Caspian Sea, but the wells lasted only a short time as they were destroyed by storms. In about 1846–1848 (authorities differ as to the exact date), Russian engineer F. N. Semyenov supervised the drilling of the first oil well on land using modern technology. By 1861, this well was producing

90 percent of all the crude oil being collected in the world (History of Drilling 2014).

Research on petroleum and its extraction were also proceeding in other parts of the world. In 1854, Polish druggist Ignacy Łukasiewicz became interested in the possibility of using crude oil as a source of kerosene as a substitute for whale oil in lamps. He had heard of a method developed by Canadian geologist Abraham Gesner for distilling crude oil to obtain kerosene, and saw the potential for a huge profit in introducing the method to his native Poland. He and his colleagues began digging wells first 50 meters, and later 150 meters, in depth, successfully collecting ever and ever purer forms of crude oil. By 1858, Łukasiewicz had constructed one of the world's first distillation plants to separate the desired kerosene from crude oil products.

Oil exploration continued apace in other parts of eastern Europe at about the same time. Drilling efforts were especially successful in Romania, where the first written production records were collected in 1857, with a reported production of 250 tonnes of crude oil for that year. Ploieşti (Ploiesti), Romania, was also the site of the world's first large oil refinery, which opened for operation in 1857 (150 Years of Oil in Romania 2014).

Explorers in North America were also becoming interested in the search for oil. The first discoveries of the fuel were made in 1858 in a region of southern Ontario around the towns of Oil Springs, Bothwell, and Petrolia (whose names all reflect their association with petroleum). The wells were extraordinarily productive, apparently having struck a vein of "inexhaustible" oil, as described by one observer of the time. Within the next decade, dozens of wells had been constructed, producing anywhere from 150 to 7,500 barrels of oil per day. Only a year after the Canadian strikes, the first successful oil well in the United States was drilled at Titusville, Pennsylvania, by "Colonel" Edwin Drake (who chose the sobriquet "colonel" to impress potential investors in his wells). Drake's discovery is sometimes claimed to be the "first well drilled solely and specifically for the purpose of finding oil," although it is hardly "the first oil well in the world," as is sometimes claimed. (For a brief history of the search for oil in the early United States, see Oil History Timeline 2014.)

For most of the nineteenth century, petroleum was a product of limited interest. In addition to some of its traditional uses, it found application primarily in the production of kerosene as a fuel for lighting systems. From 1861 to 1900, production of oil grew from 3,246,000 barrels in the decade from 1861 to 1870 to 110,593,000 barrels in the decade from 1891 to 1900. At that point, growth began to explode, reaching 274,852,000 barrels in the period from 1901 to 1910 and 610,400,000 barrels from 1911 to 1920. By the end of World War I, production was increasing by at least 300,000,000 barrels *per year*, and by the early 1920s, more than 700,000,000 barrels per year (Mineral Production 2014).

The Petroleum Century

The twentieth century might well be called the petroleum century. It was a 100-year period during which the production and consumption of crude oil mushroomed, as Tables 1.3 and 1.4 illustrate. Notice from Table 1.3 that the total amount of crude oil produced in the United States in 1860 was about a thousand barrels per day. By comparison, the average daily production of crude oil worldwide was about 1,400 barrels per day. (All worldwide figures come from The Economic Impact of Oil and Gas Extraction in New Mexico 2009, Appendix B, pages 40–41). The United States obviously accounted for a very significant fraction—just over 70 percent—of the crude oil being produced worldwide at the time.

As Table 1.3 also shows, significant growth in the petroleum market did not begin to increase in the United States until after the turn of the twentieth century. That was not quite the same

Table 1.3 Petroleum Production in the United States and the World, 1860–2013 (thousand barrels per day)

Year	Petroleum Production in the United States	Percentage of World Market
1860	1	71
1865	7	93
1870	14	88
1875	33	92
1880	72	88
1885	60	60
1890	126	60
1895	145	51
1900	174	43
1905	369	63
1910	574	63
1915	770	66
1920	1,210	64
1925	1,700	58
1930	2,460	64
1935	2,723	60
1940	4,107	70
1945	4,695	55
1950	5,407	52
1955	6,807	44
1960	7,035	33
1965	7,804	26
1970	9,637	21
1975	8,375	16
1980	8,597	14
1985	8,971	17
1990	7,355	12
1995	6,560	11
2000	5,822	8
2005	5,181	7
2010	5,471	7
2011	5,652	6
2012	6,486	6
2013	7,443	10

Sources: U.S. Field Production of Crude Oil. U.S. Energy Information Administration, http://www.eia.gov/dnav/pet/hist/LeafHandler.ashx?n=PET&s=MCRFPUS2&f=A. Accessed on June 16, 2014. Column 3 calculated from this source plus World Crude Oil Production, U.S. Energy Information Administration and "The Economic Impact of Oil and Gas Extraction in New Mexico," Arrowhead Center, New Mexico State University, September 2009, http://arrowheadcenter.nmsu.edu/sites/default/files/uploadecd/nm-oil-and-gas-impact-report-1.pdf. Accessed on June 16, 2014.

Table 1.4 Total Petroleum Consumption, United States, 1860–2013, quadrillion Btu

Year	Petroleum Consumption	Percentage of World Total[1]
1860	0.0003	
1865	0.010	
1870	0.011	
1875	0.011	
1880	0.096	
1885	0.040	
1890	0.156	
1895	0.168	
1900	0.229	
1905	0.610	
1910	1.007	
1915	1.418	
1920	2.676	
1925	4.280	
1930	5.897	
1935	5.675	
1940	7.760	
1945	10.110	
1950	13.3	
1955	17.3	
1960	19.9	
1965	23.2	37.7
1970	29.5	32.3
1975	32.7	30.0
1980	34.2	27.8
1985	30.9	31.3
1990	33.6	25.5
1995	34.5	25.2
2000	38.3	25.6
2005	40.4	23.8
2010	36.0	21.8
2011	35.3	21.4
2012	34.6	20.7
2013	35.1	20.9

[1]Reliable world data not available prior to 1965.

Sources: "Estimated Primary Energy Consumption in the United States, Selected Years, 1635–1945," U.S. Energy Information Administration, http://www.eia.gov/totalenergy/data/annual/showtext.cfm?t=ptb1601. Accessed on June 16, 2014; "History of Energy Consumption in the United States, 1775–2009," U.S. Energy Information Administration, http://www.eia.gov/todayinenergy/detail.cfm?id=10. Accessed on June 16, 2014; "Primary Energy Consumption Estimates by Source, 1949–2011," U.S. Energy Information Administration, http://www.eia.gov/totalenergy/data/annual/showtext.cfm?t=ptb0103. Accessed on June 16, 2014; "U.S. Energy Consumption in 2012 and 2013, by Energy Source," Statista, http://www.statista.com/statistics/203325/us-energy-consumption-by-source/. Accessed on June 16, 2014; Column 3 calculated from data at "Statistical Review of World Energy 2013," BP, www.bp.com. Accessed on June 16, 2014.

worldwide, where growth continued, but at a steadier pace, from 103,692 thousand barrels in 1895 to 149,137 thousand barrels in 1900 to 215,091 barrels in 1905 to 327,763 barrels in 1910. During this period, the U.S. share of the world petroleum market hovered around 50 percent, with a low of 43 percent of the world market in 1900 to a high of 63 percent during the decade of the 1910s. Throughout the century, oil production increased in the United States until about 1970, when it reached a peak of about 9.6 million barrels per day, after which it has continued to decline. (There is no evidence at this point in history as to whether or not U.S. crude oil production will again reach the 1970 peak at some time in the future, however.)

Oil consumption has followed a somewhat different pattern during the twentieth century and first decade of the twenty-first century. As Table 1.4 shows, oil consumption began to increase rapidly after the turn of the century and again in the 1980s, after which it leveled off at about 35 quadrillion Btu per year. Since the 1980s, the share of oil consumed by the United States compared to the rest of the world has very slowly decreased, partly as a result of accelerating economies in other nations of the world such as India and China.

A number of factors are responsible for the growth in petroleum consumption in the United States (and other developed nations), such as increased industrial operations and expanded production of petrochemicals. But no factor has been as important as the growth of the motor vehicle industry. In 1895, only four motor vehicles (all passenger cars) were registered in the United States. That number rose slowly to 16 in 1896, 90 in 1897, 800 in 1898, and 3,200 in 1899. Motor vehicle sales began to take off after the first decade of the twentieth century reaching the first hundred thousand in 1906 (105,900 passenger cars and 1,100 trucks) and the second hundred thousand in early 1909. Thereafter sales and registration numbers began to take off, reaching 619,500 in 1911, 1,194,262 in 1913, and 2,309,666 in 1915. (Add data from "Motor Vehicle

Registrations, 1895–1929," http://www.railsandtrails.com/AutoFacts/1930p15-100-8.jpg. Accessed on June 17, 2014). Table 1.5 shows the rate of increase in motor vehicles and passenger cars from 1920 onward.

This table makes abundantly clear the essential role that motor vehicles have played in the growth of petroleum consumption in the United States (and other developed nations) since the end of World War I. According to the most recent

Table 1.5 U.S. Motor Vehicle Registrations, 1920–2009

		Passenger Cars	
Year	Motor Vehicles[1]	Number[2]	Rate per 1,000 persons
1920	9,239	8,132	76.4
1925	20,069	17,481	150.9
1930	26,750	23,035	187.2
1935	26,546	22,568	177.4
1940	32,453	27,466	207.4
1945	31,035	25,797	193.3
1950	49,162	40,339	265.6
1955	62,689	52,145	315.9
1960	73,858	61,671	342.7
1965	90,358	75,258	388.9
1970	111,242	89,244	437.5
1975	137,913	106,706	495.2
1980	161,490	121,601	535.2
1985	177,133	127,885	537.5
1990	193,057	133,700	535.6
1995	205,427	128,387	482.2
2000	225,821	133,621	473.6
2005	241,194	136,568	n/a
2009[3]	246,283	134,880	n/a

[1]Thousand.

[2]Thousand.

[3]Latest year for which data are available.

Sources: 1920–2000: "Transportation Indicators for Motor Vehicles and Airlines: 1900 to 2001," Statistical Abstract of the United States, 2003, https://www.census.gov/statab/hist/HS-41.pdf. Accessed on June 17, 2014; 2000–2009: "State Motor Vehicle Registrations: 1990 to 2009," Statistical Abstract of the United States, 2012, http://www.census.gov/prod/2011pubs/12statab/trans.pdf. Accessed on June 17, 2014.

data available at the writing of this book, about 67.7 percent of all the petroleum consumed in the United States is used for transportation purposes (cars, trucks, trains, ships, and airplanes). That fraction is two and a half times that of the next greatest area of consumption, industrial operations (26.8 percent of all petroleum consumers in February 2014). ("Petroleum Consumption by Sector," http://www.eia.gov/totalenergy/data/monthly/pdf/sec3_18.pdf. Accessed on June 17, 2014.)

The Natural Gas Century

The beginning of the twenty-first century carried with it a crucial question for the energy industry in the United States and many other nations of the world. Petroleum consumption continued to remain high, was inching up, or was moving up rapidly in various parts of the world. Yet evidence seemed to suggest that peak oil had occurred in at least some parts of the world, and perhaps had passed that point worldwide. The challenge was, then, how the world would meet its energy demands in the twenty-first century.

A number of answers were offered for that challenge. One was conservation. Some individuals argued that the best way to bring production and consumption numbers into consonance with each other was to reduce consumption so that nations used no more oil than the world was able to produce. Others suggested a greater emphasis on alternative and renewable fuels, such as solar energy, wind power, geothermal energy, nuclear power, and the energy of waves and tides. A third recommendation with wide appeal was to explore more aggressively for new sources of petroleum. This idea has been based on the belief among many experts in the field of petroleum research that enormous reserves of oil remain undiscovered beneath Earth's surface, and that better searches will uncover the reserves humans need for many decade, and perhaps centuries, into the future. Finally, a fourth hope was that a thus-far neglected fossil fuel, natural gas, might be the key to meeting energy demands in the future.

(The eventual answer to the world's energy issues is likely to be some mix of these four options.)

The search for new petroleum reservoirs often took the form, of course, of wider and more sophisticated searches for promising petroleum sites on land. But another approach was to extend that search to offshore regions. Possibly the earliest of such offshore explorations for oil was the one mentioned earlier, in the Caspian Sea in 1803. However, most historians suggest that the first successful offshore wells were drilled in 1896 by California businessman Henry L. Williams, who owned the town of Summerland on the coast where the wells were drilled. Explorations on land in the Summerland area had already been successful, resulting in a number of productive oil wells. Williams reasoned that the oil reservoir feeding those wells might also extend below the ocean floor adjacent to the Summerland coast. He built a 300-foot-long pier from the shore out into the ocean, at the end of which a drilling rig was positioned. The experiment was so successful that five years later 22 companies had built a total of 400 producing wells on a total of 15 piers. The age of offshore drilling had begun (Offshore Petroleum History 2014).

The Summerland experience remained something of an anachronism for more than four decades before the Pure Oil Company and Superior Oil Company joined forces in 1938 to construct an even more ambitious offshore project. They hired shrimp boats to haul the parts needed to build a drilling rig about a mile offshore near Creole, Louisiana. Even at that distance, the drilling rig stood in only 14 feet of water, just barely qualifying as being an "offshore" endeavor.

Almost everyone agrees that offshore drilling truly began in 1947 with the construction of a rig called Kermac 16, the first drilling rig to be built out of sight of land. The rig was built for a consortium of Pure Oil, Superior Oil, and the Kerr-McGee Oil Industries by the Brown & Root construction company. The drilling site was still a shallow-water location, only 20 feet in depth, but it was located 10.5 miles offshore from Terrebonne Parish, Louisiana. The rig eventually produced 1.4 million barrels of oil and 307 million cubic feet of natural gas during its lifetime (Keefe 2014; Pratt 2014).

Although the general principles of drilling for oil and gas underwater are much the same on water as they are for sites on land, offshore drilling obviously poses a whole list of new problems and challenges, such as the difficulties of getting workers and equipment to and from a rig, keeping a rig in stable position above a drill hole, stabilizing a rig during extreme weather, and disposing of drilling wastes. In addition, accidents at offshore sites pose the same environmental and human safety problems they do on land, but usually in more challenging circumstances. (For an excellent discussion of offshore drilling, see Lamb 2014.) Offshore drilling also includes serious risks to human health and life as well as damage to the environment that has raised concern with many individuals and organizations over the years. (For a description of some offshore accidents, see The World's Worst Offshore Oil Rig Disasters 2014.)

In spite of its technical challenges, offshore drilling has become an important contributor to the oil production program in the United States and many other nations. Table 1.6 shows the amount of petroleum produced by two offshore well sites adjacent to the continental United States since 1981. Note that offshore wells have recently contributed about a quarter of all the oil produced by the United States.

A somewhat different pattern holds for parts of the world other than the United States. Offshore oil production globally was essentially zero prior to 1945, after which it rose to about a million barrels per day during the 1950s and early 1960s before beginning a sharp upward turn in about 1970. After that point, offshore oil production climbed to about 10 million barrels per day in 1980 and about 23 million barrels per day in 1990. During the first decade of the twenty-first century, however, offshore production leveled off at about that level, where it has remained since (Sandrea and Sandrea 2014). Interestingly enough, offshore output from shallow water wells actually

Table 1.6 Oil and Gas Production from Offshore Wells in the United States

	Source					
	Oil[1]				Natural Gas[2]	
Year	All U.S.	PADD3[3]	PADD5[4]	All Offshore	All U.S.	All Offshore
1977					21,097	3,932
1978					21,309	5,111
1979					21,883	5,603
1980					21,870	5,650
1981	8,572	719	54	773	21,587	5,693
1982	8,649	786	78	864	20,272	5,466
1983	8,688	876	84	960	18,659	4,735
1984	8,879	956	83	1,039	20,266	5,220
1985	8,971	941	82	1,023	19,607	4,632
1986	8,680	960	78	1,038	19,131	4,588
1987	8,349	892	85	977	20,140	5,078
1988	8,140	818	85	903	20,999	5,181
1989	7,613	764	91	855	21,074	5,231
1990	7,355	739	82	821	21,523	5,509
1991	7,417	799	86	785	21,750	5,308
1992	7,171	822	116	938	22,132	5,324
1993	6,847	825	139	964	22,726	5,373
1994	6,662	860	158	1,018	23,581	5,701
1995	6,560	943	196	1,139	23,744	5,432
1996	6,465	1,021	176	1,197	24,113	5,844
1997	6,452	1,129	148	1,277	24,213	5,906
1998	6,252	1,228	127	1,355	24,108	5,801
1999	5,881	1,354	109	1,463	23,823	5,689
2000	5,822	1,430	96	1,526	24,174	5,699
2001	5,801	1,536	85	1,621	24,501	5,815
2002	5,744	1,555	88	1,643	23,941	5,312
2003	5,649	1,537	81	1,618	24,119	5,216
2004	5,441	1,462	75	1,536	23,970	4,736
2005	5,181	1,279	73	1,352	23,457	3,890
2006	5,088	1,293	71	1,364	23,535	3,584
2007	5,077	1,282	67	1,349	24,664	3,477
2008	5,000	1,157	66	1,223	25,636	3,029
2009	5,353	1,562	61	1,623	26,057	3,072
2010	5,471	1,544	59	1,603	26,816	2,876

(continued)

Table 1.6 (*continued*)

	Source					
	Oil[1]				Natural Gas[2]	
Year	All U.S.	PADD3[3]	PADD5[4]	All Offshore	All U.S.	All Offshore
2011	5,652	1,316	54	1,370	28,479	2,417
2012	6,486	1,267	48	1,315	29,542	2,045
2013	7,443	1,253	51	1,304	30,171	n/a

[1]Thousand barrels per day.
[2]Billion cubic feet.
[3]Petroleum Administrative Defense District 3; Gulf of Mexico.
[4]Petroleum Administrative Defense District 5; West Coast United States.

Sources: "U.S. Field Production of Crude Oil," Energy Information Administration., http://www.eia.gov/dnav/pet/hist/LeafHandler.ashx?n=PET&s=MCRFPUS2&f=A. Accessed on June 17, 2014; "Federal Offshore–Gulf of Mexico Field Production of Crude Oil," Energy Information Administration, http://www.eia.gov/dnav/pet/hist/LeafHandler.ashx?n=PET&s=MCRFP3FM2&f=A. Accessed on June 17, 2014; "Federal Offshore PADD 5 Field Production of Crude Oil," Energy Information Administration, http://www.eia.gov/dnav/pet/hist/LeafHandler.ashx?n=PET&s=MCRFP5F2&f=A. Accessed on June 17, 2014; "U.S. Natural Gas Gross Withdrawals," Energy Information Administration, http://www.eia.gov/dnav/ng/hist/n9010us2m.htm. Accessed on June 17, 2014; "U.S. Natural Gas Gross Withdrawals Offshore," Energy Information Administration, http://www.eia.gov/dnav/ng/hist/na1090_nus_2a.htm. Accessed on June 17, 2014.

began to decrease in 2000, and stability of production figures was a result of a significant increase from deepwater wells, which produced essentially no oil before 1990. By 2010, however, those wells were producing about 5 million barrels of water per day (Sandrea and Sandrea 2014).

An interesting overall observation for most of the data shown here is that the production of oil has begun to decrease, as theories of peak oil would suggest, from all sources with the exception of deepwater drilling. The question that continues to challenge those interested in energy issues, then, is what can be done to replace the reduced levels of oil being obtained from conventional sources around the world and in the United States?

Hydraulic Fracturing

One promising answer to that question is hydraulic fracturing, a procedure for "squeezing out" crude oil and natural gas from reservoirs that had once thought to be inaccessible to drilling. Hydraulic fracturing is broadly similar to the use of a "torpedo" to open up the pockets in an oil or gas reserve. The main difference is that instead of using an explosive, such as nitroglycerine to force open these pockets, a powerful stream of liquid—almost entirely water and sand with a number of other possible additives—is forced into an existing well. The force of the stream of water is, hopefully, sufficient to open pores, cracks, and fissures, allowing oil and/or gas to flow out of the reserve and into the well.

The History of Hydraulic Fracturing

Many historians trace the origins of hydraulic fracturing to a chance observation made during the Civil War by Union officer Lieutenant Colonel Edward A. L. Roberts. During the Battle of Fredericksburg in 1862, Roberts noticed that Confederate gunners were bombarding a narrow opening in a stream that was obstructing their attack plans. Energy from the exploding shells tore apart the sides of the stream, increasing the flow of water and removing the obstruction. A few years later, returned to civilian life, Roberts considered the possibility of using a procedure similar to the one he had seen at Fredericksburg to solve a common problem associated with the operation of oil wells.

During the time that oil is being removed from a reservoir by drilling, changes take place in the reservoir. Some of the denser, more viscous components of crude oil, such as bitumen, asphalt, and waxes, begin to precipitate out of solution and to clog the pores, fractures, fissures, and other openings through which oil flows into a well. Over time, this problem becomes so severe that the flow of oil begins to slow down, and it may eventually stop entirely. Roberts reasoned that, if he could find a way to open up the blocked spaces, oil would flow more freely and wells that were no longer producing oil might once again become productive.

His solution to the problem was based on his Fredericksburg observation. He invented a device that he called a *torpedo*, a cylindrical metal container containing a small amount of explosive. The torpedo was lowered into a well and detonated. The shock produced by the explosion pushed open clogged pores, fractures, and fissures, allowing oil to flow freely once more.

Roberts's first experiments with his petroleum torpedo were a remarkable success. Some wells on which the fracturing experiment was tried increased productivity by as much as 1,200 percent within the first week following the attempt. Roberts formed a company to build and sell his torpedoes. The company became a huge success, providing Roberts with a reliable income for the rest of his life. The company remains in operation today under the name of the Otto Cupler Torpedo Company, although the technology it uses is much advanced from the one originally developed by Colonel Roberts. (Shooters—A "Fracking" History 2014; for a video of the Roberts system of clearing wells, see http://www.youtube.com/watch?v=OLdzug1C5c8.)

While torpedo-style opening of oil and gas wells remained popular over the next half century, inventors also explored other ways of opening clogged pores in such wells. At virtually the same time that Roberts made his "torpedo method" discovery, chemist Herman Frasch was awarded a patent for the use of hydrochloric acid to open fissures in an oil reservoir. (Frasch is probably better known as the inventor of the Frasch process for mining sulfur.) The principle behind Frasch's invention was that an acid could dissolve some of the rock in which oil and gas are embedded, providing larger pores through which it could flow into a well. The process worked well enough, but was bedeviled by the fact that the acid most commonly used—hydrochloric acid—also attacked the piping and other equipment used in an oil or gas well. However, inventors struggled with the challenge of finding other solutions for this problem (such as using hydrofluoric acid rather than hydrochloric acid), and the general principle of using acids as "opening agents" remains in use today. The process of adding an acid to a fracking fluid is known as *acidization* or *acidizing*. (Crowe, Masmonteil, and Thomas 1992, 24; *Dow Chemical Co. vs. Halliburton Oil Well Cementing Co.* 2014; for an excellent general discussion of acidization, see Samuel and Sengul 2003)

Another approach to opening up potential gas wells was Project Gasbuggy, a joint project of the U.S. Atomic Energy Commission, U.S. Bureau of Mines, and El Paso Natural Gas Company. Project Gasbuggy was part of a larger project, known as Project Plowshare, designed to find peacetime applications for nuclear energy which, prior to that time, had been devoted exclusively to the construction of nuclear weapons. Gasbuggy was designed to test the hypothesis that oil- and/or gas-bearing rock might be made more productive for commercial uses by exploding a nuclear device deep underground. That explosion, experts believed, would cause fracturing of rock structures, allowing oil and gas to flow more freely into wells.

Project Gasbuggy was carried out on December 10, 1967, when a 29-kiloton nuclear device was exploded 4,227 feet below ground in Rio Arriba County, New Mexico. (For comparison, the fission bomb dropped on Hiroshima had a rating of 15 kilotons.) The experiment was a success in that studies later showed that the bomb had produced extensive fracturing in the rock around the detonation site. But it had no practical value since the gas produced as a result of the fracturing was too radioactive to be used for commercial purposes. In addition, the cost of producing natural gas by this means was much greater than gas obtained by other processes.

Two other projects were conducted to obtain more information about nuclear fracturing of rock, Project Rulison, near Rulison, Colorado, on September 10, 1969, and Project Rio Blanco, near Rifle, Colorado, on May 17, 1973. By the conclusion of the latter experiment, the Department of Energy deemed nuclear fracturing to be a commercially unviable technique for gas recovery, and such experiments were discontinued (LM Sites 2014).

The most successful approach to fracturing of rock structures was based on the notion of simply increasing the pressure on those rocks, often just by adding large volumes of water to a well. One of the earliest experiments along this line was conducted in the 1930s by a process that became known as *rock busting* or *pressure parting*. One report on experiments of this type published in 1935 claimed that the output of the wells tested was increased by 20 million barrels with the use of the technique (Grebe and Stosser 1935).

Another approach that was tried was a modification of the Roberts torpedo. In 1939, inventor Ira J. McCullough received a patent for a device containing steel-covered bullets that could be shot through the lining of a well into the surrounding rock, causing fracture of the rock (Downhole Bazooka 2014).

In spite of these many beginning attempts to find ways of fracturing rock, the beginning of the modern age of hydraulic fracturing is usually traced to a series of experiments conducted in the 1940s by Riley "Floyd" Farris and his colleague, Robert Fast, at the Stanolind Oil and Gas Company. During his college years at the University of Oklahoma, Farris had concentrated on the role of cement in the construction of oil wells. As noted earlier, cement is used to provide stability to the wellbore and, hence, is an essential part of the drilling operation. Farris was puzzled by the fact that wells almost always seemed to require more cement than the amount an engineer could calculate prior to drilling. With Fast, he conducted an experiment on a trial test well and found that the reason for this discrepancy was that cement had a tendency to diffuse out of the wellbore and into the adjacent rock. But in so doing, the cement also had a tendency to break apart the rock and increase the flow of oil and gas. For Farris, the conclusion to this experiment was obvious: If flowing cement can cause fracturing in rock, why not just *intentionally* add a fluid (maybe cement, but perhaps another fluid) to produce this fracturing. With that background, Farris and Fast tried their new "hydraulic fracturing" technique on an old and worn out well in Texas that was producing about a barrel of oil per day. After treatment with a mixture of water, sand, and soap, the tired old well once again became active, producing about 50 barrels per day (Cranganu 2014, 60–61; Gold 2014, 71–74).

In 1946, Fast decided to confirm the results of the earlier fracturing experiment in what is generally regarded as the first hydraulic fracturing experiment ever conducted. Fast conducted his experiment at the Klepper #1 oil well in the Hugoton gas field in Kansas. He injected a mixture of sand and gasoline thickened with napalm to four different depths in the well. When the injection fluid was pumped out, the well began producing oil again—but only at the rate it did before treatment. The result was a failure and it seemed that, for the moment at least, the idea of hydraulic fracturing was dead (Cranganu 2014, 61; Gold 2014, 74–74).

That pessimistic conclusion was soon found to be wrong, however. Farris and Fast continued to work on their invention and in 1949, the Halliburton Oil Well Cementing Company attempted to repeat the Hugoton experiment with two other wells, one in Stephens County, Oklahoma, and the other in Archer County, Texas. In both cases, Halliburton used a process that had been patented by Stanolind in 1946 called the Hydrafrac process. That process involved the injection of "jellied gasoline" (gasoline containing napalm) into a well. In these cases, the wells began producing more productively, and the efficacy of the Hydrafrac process had been demonstrated. Evidence of its success also comes from the growing success of Halliburton itself, whose revenues increased from $57.2 million in 1949 to $69.3 million in 1950 to $92.6 million in 1951, almost entirely as a result of its success with fracking (Halliburton Oil Well Cementing Company 2014; Montgomery and Smith 2010).

Following these initial successes, hydraulic fracturing gradually grew in popularity. During the 1950s, oil companies were using the procedure on as many as 5,000 new wells each month (Montgomery and Smith 2010, 27). The value of fracking operations grew from about $2 billion in the United States in

1999 to twice that amount in 2004 to four times that amount in 2007 (Montgomery and Smith 2010, 35). At that point, fracking had become a technology entrenched in the oil and gas recovery industry.

Horizontal Drilling

A number of factors led to the growing popularity of hydraulic fracturing. In the first place, the technology described earlier had already proved itself capable of vastly increasing the amount of oil and gas that could be recovered from reservoirs that had previously been regarded as exhausted in resource or, at the very least, no longer capable of producing a fuel economically. Other factors that contributed to the growth in the use of fracking can be traced to a single individual and the company he founded, George Mitchell and Mitchell Energy.

In the 1980s, Mitchell made it his goal to find new oil and gas reservoirs wherever they might be and squeeze out the last possible barrel of cubic foot of oil or gas from those reserves. In this search, he began to explore a region that most oil companies had written off as lacking in sufficient quantities of oil and/or gas that could be recovered economically, the Bakken play that covers most of the mid-southern part of the United States. Mitchell felt certain that very large quantities of oil and gas were available in the Bakken play, although he was uncertain as to how to access those reserves.

One change he made in the fracking process was in the composition of the fluid used. Traditionally, fracking fluids had consisted of relatively thick, viscous liquids that contained gels, foams, and emulsions, such as the original gasoline and napalm mixture used in the first fracked wells. Mitchell decided to use a simpler mixture, one that consisted primarily of water (more than 90 percent of the mixture), sand, and very small amounts of other proppants. A proppant is a substance added, generally in very small proportions, that improves the efficiency of a fracking fluid for any specific application. Many of the substances previously used in fracking, those same gels, foams, gums, and acids, continued to be used in Mitchell's formations, except in much smaller proportions. Mitchell soon discovered that this new form of fracking liquid was far more effective in producing fractures than was the more viscous types of fluids used previously (Zuckerman 2014; Zuckerman's book on Mitchell's career is one of the best resources on this part of the history of hydraulic fracturing; see Zuckerman 2013).

A second major change that Mitchell introduced to the fracking process was the use of horizontal drilling. All of the discussions of well drilling in this book thus far have been about vertical drilling, a technology that involves placing a rig on Earth's surface and then drilling straight down through as many strata as necessary to reach one that contains crude oil or natural gas. Think of a layer of eight pancakes, only one of which is soaked with maple syrup. If one wishes to get to that pancake soaked with syrup, it is necessary to dig down through all the other pancakes until the syrup-soaked pancake is reached. And if, for any reason, the pancakes are not lined up exactly, it might be possible to drill right past the desired syrup-soaked pancake.

A more logical and efficient way to go about this process might be to drill into the pancake stack from the side, especially if one knows which layer contains the correct pancake. And the same is true in drilling for oil or gas. One can save all the time and expense of drilling through layers of rock that contain no oil or gas—largely a waste of time—in order to get to that one layer that does contain the desired fuel. Probably anyone who has ever drilled for oil or gas understands this concept, but that doesn't mean that it has been practiced for very long.

Interestingly enough, the first invention of a non-vertical drill for oil or gas was patented in 1891 by a dentist, John Smalley Campbell. Campbell was trying to find a way to drill into teeth to repair cavities in inconvenient locations. His patent was for a drill with a flexible shaft that could essentially go around a corner. In his patent application, Campbell noted that his invention had many other possible applications, such as drilling for oil or gas at an angle (Drilling Sideways 1993, 7).

By the 1920s, petroleum engineers had begun to think about the use of drills like those patented by Campbell for oil and gas exploration and extraction. The first such well was actually completed in 1929, near the town of Texon, Texas (Drilling Sideways 1993, 7). This well, like most of its successors, was not completely horizontal, but angled away from the vertical direction for part of its length and then extended horizontally for the last portion of its length. Such wells today are therefore said to be *directionally drilled* or *slanted drilled*, although they may also be simply called *horizontal wells*. (To see how directional drilling is done, see the animation at http://www.oerb.com/?tabid=242.)

Horizontal drilling was only moderately popular until the 1980s, largely for economic reasons. Experimentation with the technique was carried out in the United States, China, and the then-Soviet Union, but in no cases was it a major contributor to the production of crude oil and natural gas until Mitchell Energy combined the principles of directional drilling with the use of a new type of drilling fluid in a large under-explored region in the 1990s. At that point, the combination of factors worked so well for Mitchell Energy that its owner became a billionaire largely overnight, and the oil and gas community could no longer ignore the potential of this approach to the production of crude oil and natural gas (Zuckerman 2013, *passim*).

Conclusion

At the end of the twentieth century, many researchers, engineers, industrialists, politicians, and members of the general public were concerned about what they saw as the world's approaching energy crisis. Much of the existing data suggested that the demand for energy worldwide was continuing to grow, especially in developing nations with huge populations such as

China and India, while the amount of energy being produced and known energy reservoirs appeared to be decreasing. If these two trends were to continue, it was difficult to see how the world's population over the new century could continue to meet its demands for energy and, hence, to continue raising its standard of living.

At just about that time, researchers and engineers were finding ways to bring together a number of relatively new technologies—along with other technologies that had been known for decades—to create new methods for exploring and extracting the fossil fuels on which modern human society so completely depends, especially crude oil and natural gas. By the beginning of the second decade of the twenty-first century, it had become obvious that hydraulic fracturing and related technologies had the potential for reversing the apparent declines in production and reserves of oil and gas and the human race had, at the very least, been provided with a substantial amount of time in which to decide how it would deal with an oncoming energy crisis, but one that was now much farther into the future.

The problem, however, was that hydraulic fracturing, carried with it a number of problems and challenges whose solutions were not always obvious and which were serious enough to generate acrimonious debates among regulators, legislators, the legal system, petroleum engineers, environmentalists, and ordinary citizens everywhere. The nature of those problems, the choices they offer humans today, and some possible solutions suggested for dealing with these challenges are the topic of the next chapter.

References

"150 Years of Oil in Romania." 150 De Ani. http://www.150deanidepetrol.ro/history.html. Accessed on June 13, 2014.

Aliyev, Natig. "The History of Oil in Azerbaijan." Azerbaijan International. http://azer.com/aiweb/categories/magazine/22_folder/22_articles/22_historyofoil.html. Accessed on June 13, 2014.

"Ancient History Sourcebook: Greek Reports of Babylonia, Chaldea, and Assyria." Fordham University. http://www.fordham.edu/halsall/ancient/greek-babylon.asp. Accessed on June 13, 2014.

"Asphalt." http://www.ce.memphis.edu/3137/Documents/Asphalt%20History.pdf. Accessed on June 13, 2014.

Biewick, Laura R. H., Gregory L. Gunther, and Christopher C. Skinner. "USGS National Oil and Gas Assessment Online (NOGA Online) using ArcIMS." U.S. Geological Survey. [2002]. http://proceedings.esri.com/library/userconf/proc02/pap0826/p0826.htm#1. Accessed on June 7, 2014.

BP Statistical Review of World Energy, June 2013. http://www.bp.com/content/dam/bp/pdf/statistical-review/statistical_review_of_world_energy_2013.pdf. Accessed on June 6, 2014.

Burnett, H. Sterling. "How Fracking Helps Meet America's Energy Needs." National Center for Policy Analysis. http://www.ncpa.org/pub/ib132. Accessed on June 23, 2014.

Clugston, Chris. "When Is 'Global Peak Energy'?" The Oil Drum. http://www.theoildrum.com/node/2960. Accessed on June 6, 2014.

"Coal: Resources and Future Production." Energy Watch Group. March 2007. http://energywatchgroup.org/wp-content/uploads/2014/02/EWG_Report_Coal_10-07-2007ms.pdf. Accessed on June 5, 2014.

Connan, Jacques, and Odile Deschesne. 1992. "Archaeological Bitumen: Indentification, Origins and Uses of an Ancient Near Eastern Material." *MRS Proceedings.* 267: 683–720.

Cooke, Ronald R. "The Cultural Economist." http://tceconomist.blogspot.com/2014/05/peak-energy.html. Accessed on June 6, 2014.

Cranganu, Constantine. "A Short History of Hydraulic Fracturing." *Petroleum Industrial Review*. http://issuu.com/petroleumreview/docs/petroleum_review-may2014-en?Name=Value. Accessed on June 19, 2014.

Crowe, Curtis, Jacques Masmonteil, and Ron Thomas. 1992. "Trends in Matrix Acidizing." *Oilfield Review*. 4(4): 24–40. Also available at http://www.slb.com/~/media/Files/resources/oilfield_review/ors92/1092/p24_40.pdf. Accessed on June 18, 2014.

"A Crude Story: Hooray for Decay!" University of Wisconsin. http://whyfiles.org/100oil/2a.html. Accessed on June 6, 2014.

Curtis, Trisha, and Tyler Ware. "Restricting North Dakota Gas-flaring Would Delay Oil Output, Impose Costs." Oil and Gas Journal. http://www.ogj.com/articles/print/vol-110/issue-11/drilling-production/restricting-north-dakota-gas-flaring-would.html. Accessed on June 13, 2014.

Desmarest, Thierry. 2010. Quoted at ODAC, The Oil Depletion Analysis Centre. http://www.odac-info.org/newsletter/2010/01/29. Accessed on June 5, 2014.

Dow Chemical Co. v. Halliburton Oil Well Cementing Co. Halliburton Oil Well Cementing Co. v. Dow Chemical Co. https://bulk.resource.org/courts.gov/c/US/324/324.US.320.50.61.html. Accessed on June 18, 2014.

"Downhole Bazooka." American Oil & Gas Historical Society. http://aoghs.org/technology/downhole-bazooka/. Accessed on June 19, 2014.

"Drilling Sideways—A Review of Horizontal Well Technology and Its Domestic Application." April 1993. Energy Information Administration. http://www.eia.gov/pub/oil

_gas/natural_gas/analysis_publications/drilling_sideways _well_technology/pdf/tr0565.pdf. Accessed on June 19, 2014.

"The Economic Impact of Oil and Gas Extraction in New Mexico." Arrowhead Center, New Mexico State University. September 2009. http://arrowheadcenter.nmsu.edu/sites/default/files/uploadecd/nm-oil-and-gas-impact-report-1.pdf. Accessed on June 16, 2014.

Emery, Dominic, and Keith Myers, eds. 1996. *Sequence Stratigraphy*. Oxford, UK; Cambridge, MA: Blackwell Science.

Engelder, Terry, and Gary G. Lash. "Marcellus Shale Play's Vast Resource Potential Creating Stir In Appalachia." The American Oil and Gas Reporter. http://www.aogr.com/magazine/cover-story/marcellus-shale-plays-vast-resource-potential-creating-stir-in-appalachia. Accessed on June 7, 2014.

"Finding Oil and Gas from Space." Remote Sensing Tutorial. http://www.fas.org/irp/imint/docs/rst/Sect5/Sect5_5.html. Accessed on June 12, 2014.

"Glossary." U.S. Energy Information Administration. http://www.eia.gov/tools/glossary/. Accessed on June 7, 2014.

Gold, Russell. 2014. *The Boom: How Fracking Ignited the American Energy Revolution and Changed the World.* New York: Simon & Schuster.

Grebe, J. J., and S. M. Stoesser. 1935. "Increasing Crude Production 20,000,000 Barrels from Established Fields." *World Petroleum.* 6(8): 473–482.

"Haliburton Oil Well Cementing Company." Harvard Business School Baker Library. http://www.library.hbs.edu/hc/lehman/company.html?company=halliburton_oil_well _cementing_company. Accessed on June 19, 2014.

"History of Drilling." Rigs International. http://www .rigsinternational.com/our-offer/history-of-drilling/. Accessed on June 13, 2014.

"History of the Increasing Number of Motor Cars/ Automobiles." carhistory4u. http://www.carhistory4u.com/the-last-100-years/car-production/the-increasing-number-of-cars. Accessed on June 14, 2014.

"History of the World Petroleum Industry." GeoHelp Inc. http://www.geohelp.net/world.html. Accessed on June 13, 2014.

"How Many Wells in Ghawar?" Satellite O'er the Desert. http://satelliteoerthedesert.blogspot.com/2008/01/how-many-wells-in-ghawar.html. Accessed on June 7, 2014.

Keefe, Patricia. "The History of Offshore Energy." New Wave Media. http://magazines.marinelink.com/Magazines/MaritimeReporter/201404/content/history-offshore-energy-468077. Accessed on June 17, 2014.

Kopey, B. 2007. "Development of Drilling Technics from Ancient Ages to Modern Times." 12th IFToMM World Congress, Besançon (France), June 18–21, 2007. http://www.iftomm.org/iftomm/proceedings/proceedings_WorldCongress/WorldCongress07/articles/sessions/papers/A36.pdf. Accessed on June 13, 2014.

Lamb, Robert. "How Offshore Drilling Works." How Things Work. http://science.howstuffworks.com/environmental/energy/offshore-drilling2.htm. Accessed on June 17, 2014.

"LM Sites." Office of Legacy Management. U.S. Department of Energy. http://www.lm.doe.gov/default.aspx?id=120. Accessed on June 19, 2014.

"Marcellus Shale Appalachian Basin Natural Gas Play." Geology.com. http://geology.com/articles/marcellus shale.shtml. Accessed on June 7, 2014.

"Mineral Production. No. 663. Petroleum: Production, Exports, and Imports." Statistical Abstract of the United States. Fraser. http://fraser.stlouisfed.org/docs/publications/stat_abstract/pages/46114_1920-1924.pdf. Accessed on June 14, 2014.

Montgomery, Carl T., and Michael B. Smith. 2010. "Hydraulic Fracturing: History of an Enduring Technology." *JPT, Journal of Petroleum Technology*. 62(12): 26–32. Also available online at http://www.ourenergypolicy.org/wp-content/uploads/2013/07/Hydraulic.pdf. Accessed on June 19, 2014.

Mórrígan, Tariel. "Peak Energy, Climate Change, and the Collapse of Global Civilization: the Current Peak Oil Crisis." Resilience. Accessed on June 6, 2014. http://www.resilience.org/stories/2010-12-15/peak-energy-climate-change-and-collapse-global-civilization-current-peak-oil-cris

Nashawi, Ibrahim Sami, Adel Malallah, and Mohammed Al-Bisharah. 2010. "Forecasting World Crude Oil Production Using Multicyclic Hubbert Model." *Energy Fuels.* 24(3): 1788–1800.

Neo, Hui Min. "Aramco Chief Seeks to Ease Oil Production Worries." Middle East Online. http://www.middleeastonline.com/english/?id=36890. Accessed on June 6, 2014.

"Offshore Petroleum History." American Oil and Gas Historical Society. http://aoghs.org/offshore-petroleum-history/offshore-oil-history/. Accessed on June 17, 2014.

"Oil: Crude and Petroleum Products." U.S. Energy Information Administration. http://www.eia.gov/energyexplained/index.cfm?page=oil_refining. Accessed on June 14, 2014.

"Oil and Gas: History." http://grandemotte.wordpress.com/peak-oil-4-exploration-history/. Accessed on June 13, 2014.

"Oil and Natural Gas 'Reserves' — Definitions Matter." Independent Petroleum Association of America. http://oilindependents.org/oil-and-natural-gas-reserves-definitions-matter/. Accessed on June 8, 2014.

"Oil Formation." http://www.hk-phy.org/energy/power/source_phy/flash/formation_e.html. Accessed on June 6, 2014.

"Oil Formation." Science Learning. http://www.sciencelearn.org.nz/Contexts/Future-Fuels/Sci-Media/Animations-and-Interactives/Oil-formation. Accessed on June 6, 2014.

"Oil History Timeline." Oil150. http://www.oil150.com/about-oil/timeline/. Accessed on June 14, 2014.

"Petroleum Resources Management System." Society of Petroleum Engineers, American Association of Petroleum Geologists, World Petroleum Council, and Society of Petroleum Evaluation Engineers. [2007]. http://www.spe.org/industry/docs/Petroleum_Resources_Management_System_2007.pdf. Accessed on June 8, 2014.

Polo, Marco. 1936. *The Travels of Marco Polo, the Venetian.* Translated by William Marsden. London, J. M. Dent & Co.; New York, E. P. Dutton & Co. Available online at https://archive.org/stream/marcopolo00polouoft/marcopolo00polouoft_djvu.txt. Accessed on June 13, 2014.

Popescu, Adam. "A Short History of Oil: 1900 to 2010." Marketplace. http://www.marketplace.org/topics/business/maps-tough-oil/short-history-oil-1900-2010. Accessed on June 7, 2014.

Pratt, Joseph A. 2014. "*Offshore* at 60: Remembering the Creole Field." Offshore. http://digital.offshoremag.com/offshoremag/201404?pg=52#pg52. Accessed on June 17, 2014.

Samuel, Mathew, and Mahmut Sengul. 2003. "Stimulate the Flow." *Middle East & Asia Reservoir Review.* 4: 42–55. Also available at http://www.slb.com/~/media/Files/resources/mearr/num4/stimulate_flow.pdf. Accessed on June 18, 2014.

Sandrea, Rafael, and Ivan Sandrea. "Deepwater Crude Oil Output: How Large Will the Uptick Be?" Oil & Gas Journal. http://www.ogj.com/articles/print/volume-108/issue-41/exploration-development/deepwater-crude-oil-output-how-large.html. Accessed on June 18, 2014.

"The SEAB Shale Gas Production Subcommittee Ninety-Day Report," August 11, 2011. http://www.shalegas.energy.gov/resources/081111_90_day_report.pdf. Accessed on June 6, 2014.

"Shooters—A 'Fracking' History." American Oil & Gas Historical Society. http://aoghs.org/petroleum-technology/shooters-well-fracking-history/. Accessed on June 18, 2014.

Sieminski, Adam. "Outlook for U.S. Shale Oil and Gas." Argus Americas Crude Summit, January 22, 2014. http://www.eia.gov/pressroom/presentations/sieminski_01222014.pdf. Accessed on June 6, 2014.

Sorkhabi, Rasoul. "Rich Petroleum Source Rocks." GeoExPro. 2014. http://www.geoexpro.com/articles/2009/06/rich-petroleum-source-rocks. Accessed on June 6, 2014.

"Structural Traps." Petroleum Education. http://www.priweb.org/ed/pgws/systems/traps/structural/structural.html. Accessed on June 7, 2014.

"Tech Talk-Current Oil Production and the Future of Ghawar." The Oil Drum. Accessed on June 7, 2014.

"Technically Recoverable Shale Oil and Shale Gas Resources: An Assessment of 137 Shale Formations in 41 Countries Outside the United States." U.S. Energy Information Administration. June 2013. http://www.eia.gov/analysis/studies/worldshalegas/pdf/fullreport.pdf. Accessed on June 7, 2014.

"The 10 Largest Oil Deposits In The World." Business Insider Australia. http://www.businessinsider.com.au/the-ten-largest-oil-deposits-in-the-world-2011-9#ferdows-1. Accessed on June 7, 2014.

Viforeanu, Andrei, Wayne Wells, and Jay W. Hodny. 2003. "Passive Surface Geochemical Survey Leads to Dry Gas Discoveries." *World Oil.* 224(6). 53–59.

"World Bank Sees Warning Sign in Gas Flaring Increase." Press Release, July 3, 2012. http://www.worldbank.org/en/news/

pressrelease/2012/07/03/world-bank-sees-warning-sign-gas-flaring-increase. Accessed on June 13, 2014.

"The World's Worst Offshore Oil Rig Disasters." Offshore Technology.com. http://www.offshore-technology.com/features/featurethe-worlds-deadliest-offshore-oil-rig-disasters-4149812/. Accessed on June 17, 2014.

Zuckerman, Gregory. "Breakthrough: The Accidental Discovery That Revolutionized American Energy." The Atlantic. http://www.theatlantic.com/business/archive/2013/11/breakthrough-the-accidental-discovery-that-revolutionized-american-energy/281193/. Accessed on June 19, 2014.

Zuckerman, Gregory. 2013. *The Frackers: The Outrageous Inside Story of the New Billionaire Wildcatters.* New York: Portfolio Penguin.

Genie S-40
M C

Hydraulic fracturing and horizontal drilling are the exciting new breakthroughs of the twenty-first century in the petroleum industry. Companies are drilling vast numbers of new wells; reinvigorating older, supposedly worn-out wells; and making very significant profits as a result of these endeavors. According to one recent study, companies have fracked nearly 82,000 oil and gas wells since 2005, with about a quarter of those in 2012 (the last year for which data were available) alone (Ridlington and Rumpler 2013, Table 1, 20; this table is reproduced as Table 5.3 in Chapter 5 of this book). But are fracking and directional drilling really all that important in the greater scheme of our energy future?

Peak Oil and Gas: Still a Looming Threat or Only a Distant Fantasy?

American energy analysts have long been concerned about two major issues. The first is a global question: How much coal, oil, and gas remains beneath the ground that humans can reasonably expect to harvest economically? And at what point does production of these resources begin to fall behind consumer demand? That is, is there truly such a thing as "peak coal,"

Tools used to create fractures in the rock are lowered into a well at an Encana Corp. well pad near Mead, Colorado. (AP Photo/Brennan Linsley) "peak oil," and "peak gas," and, if so, at what point in history is each of these events likely to occur? This question is, of course, one of the most fundamental issues faced by humans because it raises questions about an enormous transition from the age of fossil fuels to the age of—what? Wind energy? Solar energy? Nuclear energy?

Some evidence has begun to accumulate that the development of more efficient means of extracting oil and gas—fracking and directional drilling—has significantly increased the amount of these fuels that can be recovered from previously known and unknown sources. In the United States, for example, the Energy Information Administration (EIA) estimated that there were 39.0 billion barrels of recoverable oil in the United States in 1970, the highest estimate the agency had ever announced. That estimate slowly dropped off over the years, down to 32.7 billion barrels in 1975, 29.8 billion barrels in 1980, 26.2 billion barrels in 1990, and 22.0 billion barrels in 2000. The 2008 EIA estimate fell to the lowest number since 1941 when it was announced to be 19.1 billion barrels (U.S. Crude Oil Proved Reserves 2014).

At that point, the trend in proved reserves changed direction, and the EIA estimate for 2009 reached 20.7 billion barrels, and then continued upward to 23.3 billion barrels in 2010, 26.5 billion barrels in 2011, and 30.5 billion barrels in 2012. The latest of these estimates represented an increase of about 15 percent in a single year, a rather remarkable change in predictions for a single year (U.S. Crude Oil Proved Reserves 2014).

A somewhat similar pattern was observed for natural gas reserves in the United States. For the greatest part of the period during which EIA made estimates of natural gas reserves (1979–2012), those numbers remained relatively constant at about 200 trillion cubic feet per year. But then in 2005 those estimates began to climb, reaching 248 trillion cubic feet in 2007, 318 trillion cubic feet in 2010, and 349 trillion cubic feet in 2011. (The 2012 estimate, the most recent available, dropped off somewhat to 323 trillion cubic feet. For complete data on oil and gas reserves, see Table 5.7 in Chapter 5 of this book.) (U.S. Natural Gas, Wet after Lease Separation Proved Reserves 2014)

A number of observers have noted that changes in estimates of oil and gas reserves in the United States (and elsewhere in the world) are the result almost entirely of the development of fracking and directional drilling technologies, which have revealed and opened up vast new areas for drilling that were unknown or unappreciated less than a decade ago. As H. Sterling Burnett, senior fellow at the National Center for Policy Analysis, has written:

> Just 15 years ago, analysts predicted America had only 60 years of natural gas supplies available at then current rates of use. Today, natural gas consumption is much higher, and fracking has increased estimated reserves to 100 years or more. (Burnett 2014)

As one might expect, estimates of oil and gas reserves in the United States and elsewhere are subjects of some disagreement. Various agencies and companies use different methods and criteria to decide what is and what is not a "proved reserve," so not everyone agrees with EIA estimates. The unreliability of such estimates was demonstrated in early 2014 when the EIA decided to change rather drastically its estimates of resources at one major U.S. oil and gas field, the Monterey Play in central California. In its 2011 report, "Review of Emerging Resources: U.S. Shale Gas and Shale Oil Plays," the agency had estimated that the Monterey Play contained an estimated 15.4 billion barrels of tight oil, by far the largest reservoir of such oil in the United States at the time. Based on that estimate, the Monterey Play was thought to be responsible for 64 percent of the total oil shale reserves in the United States (Review of Emerging Resources 2011, 4; for industry reaction to this news, see Redden 2012).

Only three years later, however, the EIA changed its mind about this estimate. In reconsidering the data collected for the agency by INTEK, Inc., the EIA concluded that a more realistic estimate for oil reserves available from Monterey was 600 million barrels, 96 percent less than its original estimate. A number of observers took note of this reassessment, pointing out, among other things, how difficult it is to make reliable estimates of fuel reserves and, therefore, how undependable predictions for the nation's energy future are likely to be. (See, for example, Murphy and Murphy 2014; the original story on the EIA reevaluation is at Sahagun 2014.)

Various observers had different reactions to this apparently devastating news. Industry spokespersons and advocates, for example, often took a surprisingly moderate view of the news, suggesting that concerns about the "loss" of oil and gas reserves were an overreaction to the realities on the ground. Tim Worstall, a contributor to the business magazine *Forbes*, for example, said that the actual *amount* of oil and gas in the ground was not really the issue. "After all," he wrote after the EIA announcement, "there's still exactly the same amount of oil there as there was before the announcement. All we've been told is that it's going to be more difficult to get it out" (Worstall 2014). Amy Myers Jaffe, executive director of energy and sustainability at the University of California at Davis, suggested that the EIA announcement did not take into consideration the inventiveness of oil and gas companies. "The academic geologists said the same thing about the Barnett Shale," she said. "They said it wasn't recoverable and it wasn't going to produce. That didn't turn out to be right. . . . Some of those companies are not going to give up. The first companies that went into North Dakota had problems too. It's very early to be making a definitive judgment" (Green 2014).

Advocates of a peak energy position had a very different view of the EIA report. They saw the reduced estimates of oil and gas reserves in the Monterey Play as confirmation of their view that the era of recoverable oil and gas was truly coming to an end. As one proponent of that position, J. David Hughes, suggested, "The oil [in the Monterey Play] had always been a statistical fantasy. . . . Left out of all the hoopla," he went on, "was the fact that the EIA's estimate was little more than a back-of-the-envelope calculation" (Monterey Shale Downgraded 2014; also see Hughes' report, Drilling California, 2013).

The EIA's own chief investigator into the Monterey field explained how initial estimates could have been so wildly off target. "From the information we've been able to gather," he said, "we've not seen evidence that oil extraction in this area is very productive using techniques like fracking. . . . Our oil production estimates combined with a dearth of knowledge about geological differences among the oil fields led to erroneous predictions and estimates" (Sahagun 2014).

In some ways, the most perceptive view of this situation may be that of specialists like Ryan Carlyle, a hydraulics engineer, who wrote in a 2013 issue of *Forbes* magazine that "reserves statistics . . . are consistently wrong," although, he went on, "for complex reasons" (Carlyle 2014). In sum, Carlyle wrote:

> Corporate reserve estimates are too low by an unknown amount, and national reserve estimates are too high by an unknown amount. And no one knows how much oil is still undiscovered—in the Gulf of Mexico alone, about 1 billion barrels of new oil have been found each year for 25 years (except very recently due to moratorium slowdowns). Shale oil fracking outside the US could add enormous resources, and simple technology improvements in gas-to-liquid conversion technology could create massive gains in production on top of that (Carlyle 2014)

All of which means that, even under the best of circumstances, we need to take energy reserves data with a grain (or shaker-full) of salt. Perhaps the only thing that oil and gas specialists from a variety of philosophical positions can agree on is that fracking and directional drilling are the factors that have been responsible for the explosion in the growth of oil and gas reserves and production over the past decade. As Christof

Rühl, chief economist at BP, has written, the massive increase in oil and gas production in the 2010s has been the result of these two new technologies producing "the world's largest increase in oil as well as gas production . . . [as well as] the biggest oil production increase in the history of the US—a country, mind you, which has produced oil since 1859" (Rühl 2014).

The Global State of Affairs

Given this word of caution, one might ask what the status of gas and oil reserves are in countries other than the United States, and to what extent fracking and directional drilling have influenced these estimates. In January 2014, the EIA released a report summarizing the estimated shale oil and shale gas reserves for a number of countries worldwide. The report indicated that the nation with the largest shale oil reserves in the world was Russia, with reserves estimated at 75 billion barrels (bbl) of shale oil, followed by the United States (58 bbl), China (32 bbl), and Argentina (27 bbl). The nation with the largest shale gas reserves according to the EIA was China, with 1,115 trillion cubic feet (tcf) of reserves, followed by Argentina (802 tcf), Algeria (707 tcf), and the United States (665 tcf) (Shale Oil and Shale Gas Resources Are Globally Abundant 2014; for a complete table for these resources, see Table 5.8 in Chapter 5 of this book).

The striking point about these data, however, is that fracking has yet to be introduced into almost any nation in the world other than the United States and Canada. According to the EIA, 39 percent of the natural gas produced in the United States in 2012 came from fracked shale reservoirs. The only other nation in the world that produced significant amounts of shale gas was Canada, where 15 percent of the nation's total natural gas production came from fracked wells. China reported the beginnings of a fracking industry, but less than 1 percent of all the natural gas produced in the country came from fracked wells (North America Leads the World in

Production of Shale Gas 2014). A number of other nations were experimenting with fracking of shale formations, but none had yet produced commercial quantities of oil or gas by the process. (For a detailed discussion of the progress in fracking in 41 countries of the world, see Technically Recoverable Shale Oil and Shale Gas Resources: An Assessment of 137 Shale Formations in 41 Countries Outside the United States 2014. For a graphical representation of the global imbalance of short-term crude oil supplies available through shale resources, see BP Energy Outlook 2030 2013, "Tight Oil Output," page 34.)

For example, YPF, Argentina's national oil company, signed an agreement in September 2012 with Chevron to begin exploring the use of fracking in its oil- and gas-rich Vaca Muerta region (Kelly 2014). At about the same time, early fracking experiments in Australia's Queensland state led to violent protests about use of the technology in attempts to recover abundant oil and gas resources in the region (at just about the same time that the nearby state of Victoria banned fracking) (Australia's Victoria State Bans Coal Seam Gas "Fracking" 2014; Gas Goes Boom 2012). Indeed decisions to ban the use of fracking in gas and oil recovery appear to be about as common as decisions to use the technology to begin recovering a nation's proved reserves of shale oil and gas. As of late 2014, fracking had been banned, at least on a temporary basis, in a number of countries, including Bulgaria, the Czech Republic, France, Germany, Romania, and South Africa; many states and provinces, such as Cantabria (Spain), Fribourg (Switzerland), Newfoundland (Canada), and Vermont (United States); as well as a host of town, cities, and other municipalities (9 Countries or Regions That Ban Fracking 2014; List of Bans Worldwide 2014; for an excellent overview of the possible future of fracking worldwide, see Shale Gas: A Global Perspective 2011 and Shale Oil: The Next Revolution 2013).

Global perspectives on the future of shale oil and gas are particularly interesting because of overall patterns of oil and gas production worldwide. In a recent report, the EIA noted that world production of crude oil, not including U.S. shale oil, has been dropping off since 2012 and has reached levels not seen since 2005 (see the graph at Mushalik 2014, Figure 2, for an illustration of this trend). When production from U.S. shale reservoirs is included, however, worldwide production of crude oil is increasing by about 5 percent per year. Although the world is at an early stage in this trend, it seems likely that harvesting of shale oil (and shale gas) resources may be the only way that worldwide oil and gas production will continue to increase in the near future.

Energy Independence for the United States

The second question for energy analysts who focus primarily on the United States is the issue of energy independence. Over at least the past century, Americans have come to depend more and more heavily on fossil fuel resources from other nations, primarily but not exclusively Middle East nations such as Saudi Arabia, Kuwait, and the United Arab Emirates. But this dependence has had all manner of economic, political, military, social, and other ramifications. For this reason, every president since Richard Nixon has promoted some form of "energy independence" for the United States, although each president has suggested a different route to this goal, such as increased conservation, greater reliance on America's own natural resources, and/or more aggressive development of alternative and renewable resources (Gottesdiener 2014).

Data suggest that these efforts were largely unsuccessful until very recently. As Table 2.1 shows, the percentage of petroleum and natural gas imported from outside the United States steadily rose in actual amounts and in percentage of domestic consumption from the 1950s until about 2007. After that point, imports for both products began to drop off dramatically, largely as a result of increased domestic production of both fuels. In fact, officials at the International Energy Agency

Table 2.1 U.S. Net Import of Petroleum and Natural Gas, Historical Trends

Year	Oil[1]	Percent[3]	Natural Gas[2]	Percent
1950	318	8.4	0	0
1955	880	10.4	11	0.13
1960	1,613	16.5	156	1.30
1965	2,281	19.8	456	2.98
1970	3,161	21.5	821	3.88
1975	5,846	35.8	953	4.88
1980	6,365	37.3	985	4.96
1985	4,286	27.2	950	5.50
1990	7,161	42.0	1,532	8.00
1995	7,886	44.5	2,841	12.79
2000	10,419	52.9	3,782	16.21
2005	12,549	50.3	4,341	19.72
2006	12,391	59.9	4,186	19.29
2007	12,027	58.1	4,608	19.94
2008	11,090	56.9	3,984	17.11
2009	9,654	51.4	3,751	16.37
2010	9,435	49.2	3,741	15.53
2011	8,935	47.3	3,469	14.17
2012	8,527	48.3	3,138	12.29
2013	7,719	40.9	2,883	11.07

[1]Thousand barrels per day.
[2]Billion cubic feet per year.
[3]Percentage import of domestic consumption.

Sources: "Petroleum and Other Liquids Overview, Selected Years, 1949–2011," U.S. Energy Information Administration, http://www.eia.gov/totalenergy/data/annual/pdf/sec5.pdf. Accessed on June 22, 2014; "Natural Gas Trade by Country," U.S. Energy Information Administration, http://www.eia.gov/totalenergy/data/monthly/pdf/sec4_4.pdf. Accessed on June 22, 2014; "U.S. Natural Gas Total Consumption," U.S. Energy Information Administration, http://www.eia.gov/dnav/ng/hist/n9140us2a.htm. Accessed on June 22, 2014.

announced in late 2013 that the United States would finally surpass both Saudi Arabia and Russia, the world's two leading oil producers in that year, by 2017 at the very least, and probably as early as 2015 (Smith 2014).

Whether the United States is at a point of achieving energy independence and, if so, what effects that will have on this country and the rest of the world are still points of serious debate. On the optimistic side (from one perspective), the evidence seems clear that the vast shale oil and gas resources that have been discovered in the United States, not even considering even greater reserves that may yet to be found, almost guarantee that the United States will continue to produce a larger share of the fuels it needs to power our economy. Such a change could truly be revolutionary on both the domestic and international level. Energy independence would mean, of course, that the United States would no longer have to import most or even any of the oil and gas it needs to run its economy in the future. That change could have a number of political, economic, social, and other implications for American society.

For example, some studies suggest that the boom in shale oil and gas has already resulted in an increase of 2.1 million jobs in the United States since 2012, and that number could rise by more than a million by 2020. An improving economy made possible by growth in the oil and gas industry could also mean an additional $74 billion income for the federal government in federal and state revenues along with average savings of $1,200 per American household because of lower energy prices (Yergin 2014). Much of the economic benefit to the United States would come from a dramatic decrease in payments to oil-producing countries, such as Saudi Arabia and Venezuela, once energy independence is achieved. In 2013, for example, the United States spent $300 billion for imported oil, a savings under energy independence that would result in a 2 percent annual economic growth for the country (Anderson 2014). Even if this estimate is too high, it does not detract from the fact that the United States would no longer be devoting the largest fraction of its trade balance deficit to importing oil, as is the case today.

Of course, energy independence would have profound effects on other nations around the world in addition to its transformation of U.S. society. The countries most likely to suffer from U.S. energy independence are those for whom oil and gas exports constitute a significant portion of the nations' economies. For Ecuador and Colombia, as two examples, oil and gas exports to the United States currently account for 8 and 7 percent, respectively, of national income. Loss of U.S. dollars might well be catastrophic for such nations (Anderson 2014).

Other nations not in the business of exporting oil and gas would probably also suffer from U.S. energy independence. As oil and gas become less expensive in the United States (because they can now be obtained domestically), manufacturing processes are also likely to become less expensive. Such a trend would probably lead to U.S. companies that had once fled the United States because of high production costs returning to their home base. In such a case, many countries around the world would see a readjustment of their own economies necessitated by the loss of U.S. multinational corporations (Anderson 2014; Menenberg 2012).

The possible political consequences of an energy independent United States are also profound. For many decades, United States foreign policy has been dictated to at least some extent by the necessity of placating nations on which it depends for a significant portion of its oil and gas. The United States could hardly afford to antagonize Saudi Arabia and other Middle East nations when so much of its economy depends on fuels imported from those nations. Energy independence would mean that it might no longer make any difference one way or another as to whether OPEC members approved of U.S. foreign policy decisions because the United States would no longer be dependent on the natural resources imported from those countries (see, for example, US "Energy Independence": Geopolitical Consequences 2013).

In spite of the apparent appeal of energy independence in the United States, a number of observers argue that such a condition is not at hand, and not likely to occur in the foreseeable future, nor many of them add, should it be allowed to take place. Perhaps the most comprehensive expression of this view comes from Robert Bryce, a fellow at the Institute for Energy Research. In his 2008 book, *Gusher of Lies*, Bryce starts out by saying that "energy independence is hogwash" (Bryce 2008, 7).

Bryce then goes on to describe in great detail the reasons that energy independence is unlikely to occur for technical reasons (the book was written just before fracking operations became so widely popular in the United States) and, more importantly, why such a condition "makes no sense" from an economic, political, military, or environmental standpoint. "Worse yet," he goes on, "the inane obsession with the idea of energy independence is preventing the U.S. from having an honest and effective discussion about the energy challenges it now faces" (Bryce 2008, 7). In a nutshell, Bryce's argument is based on the reality that the world today operates less as a group of individual nations, each with its own economic, political, social, military, environmental, and other concerns in isolation from the rest of the world, and more as an interdependent whole in which decisions made by any one country have far-reaching effects on many, most, or all other countries on the planet. Thus, America's energy future cannot be viewed just in terms of what is good for the United States, but as to how it will contribute to a well-functioning world community.

Bryce is by no means the only person to argue against the inevitability and/or advisability of an energy independent United States. In a 2012 article on Salon, for example, Michael T. Klare, professor of peace and world security studies at Hampshire College in Amherst, Massachusetts, points out that energy independence for the United States is likely to come about only through the use of what he calls "extreme" technology. Conventional sources of oil and gas continued to be depleted, he says, so that energy independence will occur only as energy companies "push the envelope," using ever more expensive and complex technologies (such as fracking and directional drilling) to extract tightly held oil and gas in shale and other "tight" formations. He suggests the use of these "extreme measures" will almost inevitably set off a dangerous chain of events that he describes as "extreme energy = extreme methods = extreme disasters = extreme opposition." He concludes that "were such a publicized golden age [of energy independence] to come about, we would be burning vast quantities of the dirtiest energy on the planet with truly disastrous consequences" (Klare 2012).

A particularly interesting analysis of the claims for energy independence from a business standpoint was offered in a 2013 report by financial consultant Deborah Rogers. In the conclusion to her report, Rogers offered a number of reasons to be suspicious about claims of energy independence as being "just around the corner," such as:

- Wall Street (business investors) have been promoting shale gas drilling so aggressively that the price of gas has dropped below production costs, which has produced very large profits for the Street through acquisitions, mergers, and other non-productive activities.
- Claims for energy independence have been based on estimates of shale oil and gas reserves that are anywhere from 100 percent to 500 percent overestimates.
- The rapid drop-off in production of shale oil and gas appears to be very similar to that observed in the past for conventional oil and gas resources.
- The oil and gas industry has already begun to pull back from new investments in exploration for and extraction of shale oil and gas because of a number of factors, such as the high cost of production and popular objections to the practice (Rogers 2013).

Yet another expression of doubt about the triumph of energy independence in the United States has come from Michael Levi, David M. Rubenstein senior fellow for energy and the environment at the Council on Foreign Relations and author of a recent book on the topic, *The Power Surge: Energy, Opportunity, and the Battle for America's Future* (Levi 2013b). Levi points out the irony of Americans glorying in the possibility of energy independence when, for the past several decades, policymakers have pursued an initiative to make sure that the world's nations would have ever more integrated economies. The fact that the United States would soon be able to produce all the oil and gas it needs would not, he says, change that reality. The U.S. economy would still be intimately tied to that of other nations, and it could not begin to act as if it were an isolated outpost in the world of petroleum economics (for a brief summary of Levi's views see Levi 2013a).

Benefits of Fracking

Anyone who chooses a newspaper or magazine article to read about fracking, or who searches for the term on the Internet, is likely to encounter a number of reasons to be concerned about the technology, reasons to impose severe restrictions and regulations on fracking or even to ban the procedure outright. As noted earlier, a very large number of governmental units, from national governments to local municipalities, have already taken this step. Before considering the objections to fracking, however, it is worthwhile to note that hydraulic fracturing has a number of benefits for both society at large and for individual citizens.

Perhaps the most obvious benefit is one that has already been mentioned and discussed at length earlier: Fracking has made possible a very significant increase in the amount of oil and gas produced in the United States and, to a lesser extent, Canada. It also holds out the promise of producing a similar effect in a number of countries around the world. But a number of other benefits can also be identified.

Direct Economic Benefits

Direct economic benefits are obvious improvements in measures such as average household income, number of jobs in an area, increase in tax revenues, or other benefits that arise from some policy decision or action. For example, when a company uses hydraulic fracturing in a county, it produces economic improvements that everyone can see and that can be easily measured, such as more jobs that might pay better than the jobs they replace or that did not exist before the company began fracking, as well as the additional income earned by businesses in the region and increased tax revenue to municipalities in the county.

A number of studies have been done that attempt to assess the direct economic benefits that accrue to a community and a region as a result of new drilling and extraction technologies that were not previously used in the area. For example, a study reported in 2011 attempted to measure the direct economic benefits that would come to the state of New York, where fracking was banned in a number of towns, cities, and counties at the time, if such bans were lifted. That study found that economic output in the state would increase by $10.4 billion, between 15,000 and 18,000 jobs would be created in the so-called Southern Tier of New York and in western New York, an additional 75,000 to 90,000 jobs would be created if the area cleared for fracking were to be expanded across the state, and local municipalities would see an increase in taxes of $1.4 billion (Considine, Watson, and Considine 2011). (It perhaps goes without saying that almost any study of this nature is likely to have its critics who find fault with methodology and/or results, critiques that will not be reviewed here. But see, as an example, Heinberg 2014.)

The American Gas Association (AGA) has also made its own estimates as to the direct economic benefits of shale gas extraction projects. In it 2014 edition of the organization's *Playbook*, AGA estimates that projects for the extraction of shale gas in the United States will generate 1.4 million new jobs, $197 in gross domestic profit, and $50 billion in additional tax revenues, just by the year 2015 (Fueling the Future with Natural Gas: Bringing It Home 2014).

Arguably the gold standard for direct economic benefits from hydraulic fracturing operations in the United States is the state of North Dakota, which has experienced what can only be called a "boom" in oil and gas production in less than a decade. A number of studies have been conducted on the economic impacts of fracking operations on the state with findings that include:

- The lowest unemployment rate of any state in the United States, 2.6 percent in early 2014. The next lowest rate was South Dakota (3.6 percent), and the national average was 6.7 percent (2013 State Unemployment Rates 2014).
- Independent economists estimated the annual employment gain in North Dakota would be 1.72 percent, exceeding the next two states in that category, Florida (1.57 percent) and Texas (1.56 percent) (Brown 2014).
- Individual income in the state has also increased significantly since fracking operations began in the late 2000s, with many residents receiving anywhere from $50,000 to $100,000 a month in leases and royalties for fracking conducted on the land they own. According to one estimate, the fracking boom is producing about 2,000 new millionaires every year in North Dakota (Bailey 2014).
- The state of North Dakota can expect to receive average tax revenues in the amount of $15,450,151 over the first 36 months of operation of each oil well in the state, of which 89 percent goes to the state and the remaining 11 percent goes to local municipalities. Current estimates are that 48,000 wells will need to be drilled and fracked over the next two decades to extract all of the oil in the Bakken formation that underlies the state. If that estimate is correct, the state should reach about $720 billion over that period in tax revenues (Benefitting from Unconventional Oil 2012).

An exhaustive report on the potential benefits of shale oil and gas was prepared under the sponsorship of a consortium of energy-related associations including the American

Chemistry Council, America's Natural Gas Alliance, the American Petroleum Institute, the Fertilizer Institute, the U.S. Chamber of Commerce Institute for 21st Century Energy, the National Association of Manufacturers, the Natural Gas Supply Association, Rio Tinto, and the Society of the Plastics Industry. The report is entitled "America's New Energy Future" and appears in three volumes, "National Economic Contributions" (http://www.ihs.com/images/Americas-New-Energy-Future-National-Main.pdf), "State Economic Contributions" (http://www.ihs.com/images/Americas-New-Energy-Future-State-Main-Dec-2012.pdf), and "A Manufacturing Renaissance—Main Report" (http://www.ihs.com/images/Americas-New-Energy-Future-Mfg-Renaissance-Main-Report-Sept13.pdf, all files accessed on June 25, 2014).

Indirect Economic Benefits

Indirect economic benefits are those that occur in some area other than the primary area of discussion or concern, in the case of hydraulic fracturing, at some point other than within or around the operation of fracking itself. Such benefits are sometimes said to occur "upstream" or "downstream" from the main activity—that is, before the activity (fracking) actually occurs, or as a result of that activity's having been carried out. For example, one indirect economic benefit of fracking is often a reduction in the price of electricity. As larger quantities of natural gas are produced as a result of hydraulic fracturing, the price of the product (gas) has a tendency to fall. The cost of electricity that is produced by the combustion of natural gas, then, is also likely to fall, and residential, municipal, and industrial consumers are all likely to see reductions in their electrical bills as a long-term effect of fracturing operations. (For an extended discussion of such indirect benefits, see Fact Sheet 2014; Hassett 2013.)

One recent study attempted to estimate the total direct, indirect, and induced effects of developing a shale gas field in

Illinois. In this study, the author added *induced* effects to his analysis as those "changes in regional household spending patterns caused by changes in household income generated from the direct and indirect effects" (Loomis 2012, 11). He found that the number of new jobs created by fracking operations under the conditions studied ranged from 602 to 5,422 as a result of direct effects, 111 to 999 under indirect effects, and 320 to 2,881 under induced effects, all estimates depending on the number and type of wells that might be drilled in the New Albany field (Loomis 2012, 13). His estimates for direct, indirect, and induced labor income under a variety of scenarios ranged from $29.9 million to $269.3 million; $8.7 million to $78.1 million; and $15.2 million to $137.2 million, respectively (Loomis 2012, 13).

Environmental Benefits

One of the indirect benefits most commonly touted for fracking is the potential bonuses the technology may offer to the natural environment, especially in the area of global climate change. Prior to the rise in popularity of fracking, a number of environmentalists argued that converting U.S. industrial operations from coal to natural gas would provide a significant benefit to the global environment because the combustion of natural gas produces about half the carbon dioxide as the combustion of comparable amounts of coal. As a possible "way station" toward the elimination of fossil fuels entirely, then, some experts recommended converting coal-fired power plants to natural gas as soon as possible. In one important study of the replacement of coal by natural gas, researchers suggested converting the construction of 28 gigawatts of natural gas power plants annually in place of the same amount of coal-powered power plants over the next half century, resulting in a savings of one gigaton of carbon emissions each year (Pacala and Socolow 2004, 17).

This theme—that using natural gas rather than coal for industrial processes reduces greenhouse gas emissions—has been repeated in any number of studies in the past decade. It has, in fact, become an important talking point by proponents of fracking who now consistently argue that fracking operations have become an important factor in the world's fight against global climate change. (For a summary of relevant research, with links to studies supporting this point of view, see Everley 2013b.)

Somewhat ironically, however, a number of environmentalists who once supported the use of natural gas as a way of reducing carbon dioxide emissions, especially prior to the development of hydraulic fracturing, have reversed their positions in the last few years. The reason for this change of heart has been the recognition that, while natural gas is a cleaner fuel to use for industrial operations, it still contributes to climate change in a different context, namely during the process of extraction. This realization has come about as research shows that the amount of natural gas leaking from a well is not insignificant. In fact, some studies show that anywhere from 3.6 to 7.9 percent of the methane obtained from shale gas wells leaks into the air, eventually escaping into Earth's atmosphere (Howarth, Santoro, and Ingraffea 2011, 679). Although it remains in the atmosphere for a much shorter time than carbon dioxide, it has more than 20 times the ability to trap heat as does carbon dioxide and is, therefore, at least as significant as carbon dioxide in contributing to global climate change (Overview of Greenhouse Gases 2014).

A somewhat middle ground taken by some observers is that shale gas certainly has a number of environmental disadvantages (more on this point later), but it still represents a great step forward in efforts to save the natural environment and should be supported by anyone who is concerned about that issue. As one document with this orientation begins, "Environmentalists who oppose the development of shale gas and fracking are making a tragic mistake." Even though shale gas does pose some environmental issues, the document goes on, "[t]hese concerns are either largely false or can be addressed by appropriate regulation" (Muller and Muller 2013).

The debate over the contribution of fracking to global climate change is by no means over. It goes back and forth with the "he said, she said" character of many issues related to the use of hydraulic fracturing, and is likely to do so well into the future. (For an excellent summary of the most recent information available on this topic, as of late 2014, see Brandt et al. 2014 and Golden 2014.)

Opposition to Fracking

As with most controversial social issues, hydraulic fracturing has been the subject of a number of public opinion polls attempting to assess individuals' familiarity with the topic and their attitudes in favor of or in opposition to the practice. In one of the most comprehensive of those polls, the Pew Research Center for the People and the Press found that only a quarter (26 percent) of respondents had heard "a lot" about fracking, while a third (37 percent) said they had heard "a little," and another third (37 percent), "nothing at all." Those percentages remained constant across political lines among Democrats, Independents, and Republicans. Of those who had heard "a lot" about fracking, just over half (52 percent) favored the practice, although that percentage varied significantly across party affiliations. About three quarters (73 percent) of all Republicans favored the use of fracking in oil and gas extraction, while only a third of Democrats (33 percent) and half of Independents (54 percent) held that position (As Gas Prices Pinch, Support for Oil and Gas Production Grows 2012).

A number of public opinion polls have focused on specific states or regions. For example, researchers at the University of Michigan queried residents of Michigan and Pennsylvania about their knowledge of and opinions about fracking in May 2013. This survey is of special interest because researchers asked a number of questions about specific aspects of fracking, such as its potential risks and benefits. As with the Pew study, the Michigan study found that less than half of respondents had even heard about fracking, with 40 percent in Michigan and 46 percent in Pennsylvania saying that they had "heard a lot" about the practice, and virtually the same numbers (42 percent and 40 percent, respectively) having heard "a little." (This study is especially interesting because Pennsylvania is one of the most heavily "fracked" states in the United States, while Michigan has no more than a handful of such wells.)

When asked their views on fracking, Michigan and Pennsylvania respondents had roughly similar views about the practice with 54 percent of Michiganders saying that they supported fracking "strongly" or "somewhat," compared to 49 percent of Pennsylvanians who expressed the same view. In something of a contrast, however, about a third of both groups (31 percent in both cases) had a negative attitude toward the term *fracking* itself, while 45 percent (again the same in both groups) held negative attitudes about the term.

In responding to more detailed questions about fracking, both Michigan and Pennsylvania respondents also held very similar views, with just over half of both groups saying that they thought fracking was associated with more benefits than problems both now and in the future (a difference of less than 3 percent in all forms of the question between states). And all respondents pointed to two specific benefits—energy independence and a source of jobs—as being the major benefits of fracking in their states. The greatest difference between the two groups was in their perceptions of the most serious possible disadvantage of fracking, with Pennsylvanians naming contamination of water supplies as the most troubling problem (34 percent of respondents) and Michiganders saying that they didn't know of any really serious problem (25 percent). These numbers obviously reflect to some extent the experience that residents of the two states have already had (or not had) with respect to hydraulic fracturing (Brown et al. 2013).

Problems Associated with Hydraulic Fracturing

The one thing that a person is probably most likely to think about when considering the possible problems associated with hydraulic fracturing is water:

- The 4.4 million gallons of water needed to frack the average oil or gas well (How Much Water Does It Take to Frack a Well? 2013);
- The alternative residential, commercial, industrial, and other uses to which that water could otherwise be put;
- The complications involved in transporting that water from its source to the fracking site;
- The chemicals that must be added to that water to maximize its efficiency in the process of fracking;
- The disturbance that water creates in rock structures into which it is injected, disturbance that may result in earthquakes and other seismic events;
- The production of additional amounts of water originally stored underground, but released by the fracking process;
- The contamination of the fracking water by the fracking process itself;
- The ultimate fate of the original fracking water along with the newly released water, especially upon its return to the surface; along with the myriad effects on human health and the natural environment caused by the water used in the fracking process and all of the other by-products of that technology. What, then, can be said about these common problems often associated with the fracking of oil and gas wells? Here are some of the most common answers to that question.

Water Use

Critics of fracking often point out that the technology wastes enormous amounts of water. The primary objection is not that fracking uses a lot of water—many industrial operations do so—but that the water used for that purpose is essentially lost for other possible uses. A nuclear power plant, for example, uses very large quantities of water to cool the reactor core. But that water is stored for a period of time in cooling ponds and then returned to lakes and rivers, from which it can then be extracted for other purposes. But water used for fracking can generally not be recycled in the same way. It may remain buried underground; it may be pumped into empty wells as permanent storage sites; it may be recycled in the fracking of other wells; or it may be treated in wastewater plants and then returned to rivers and streams. The water is very unlikely to end up in a drinking water reservoir, in a farmer's irrigation ditch, or on its way to an industrial plant (Easton 2014; Hansen 2014).

A number of studies have attempted to obtain more specific information about the fate of water used in fracking operations, often with only limited success. In one such study of water use in fracking operations in Pennsylvania and West Virginia, for example, researchers noted that one challenge they encountered was that state reporting requirements were such that only very limited conclusions could be drawn about the way in which fracking water was used and then disposed of (Hansen, Mulvaney, and Betcher 2013). These researchers were, nonetheless, able to reach some qualified conclusions:

- An average of 4.3–5 million gallons of water is used in the fracking of wells in the region.
- About 80 percent of the water used for fracking is taken directly from rivers and streams.
- Between 6 and 8 percent of the water used in fracking is recovered; the balance remains underground, unavailable for future use of any kind.
- Two-thirds of the flowback water from West Virginia wells is unaccounted for because state regulations do not require full reporting of this information.

- More than half of the wastewater from Pennsylvania wells is treated and returned to the environment, while essentially none of the West Virginia wastewater goes through this process (Hansen, Mulvaney, and Betcher 2013; for detailed discussions of these points, see sections 4 and 5 of the report, pages 15–38).

Regardless of the numerous studies that have been conducted, a number of ordinary citizens and the organizations to which they belong are worried about the massive use of water for fracking. That concern is only made more serious if and when drought conditions develop in a region where fracking is occurring or is planned. In late 2012, for example, citizens of Carroll County, Ohio, expressed their growing concerns that the water be withdrawn from the region's water table would result in a deficiency of water for residential, industrial, commercial, and other applications. Even though some of those citizens were accepting payment for water being withdrawn from their own ponds, the consensus view seemed to be that better controls were needed to make sure that the county did not run out of water as oil and gas companies continued to drain the country's natural sources of water (Hunt 2012).

The oil and gas industry's primary response to the concern about water use in the fracking process is that critics should keep the issue in perspective. Every form of energy generation requires the use of water—usually very large amounts of water—and fracking is no exception. The major difference, they say, is that fracking requires *much less* water use than do most other forms of energy generation. In a presentation made to the Ground Water Protection Council in 2009, for example, Chesapeake Energy displayed a table showing the amount of water used for a variety of energy generation technologies, in gallons of water used per million Btu of energy. The table showed that the extraction of shale natural gas requires anywhere from 0.60 to 1.80 gallons per million Btu (gal/MMBtu), compared to 1 to 3 gal/MMBtu for conventional natural gas, 2 to 32 gal/MMBtu for coal (depending on method of transport), 8 to 14 gal/MMBtu for nuclear power, 8 to 20 gal/MMBtu for conventional oil, 22 to 26 gal/MMBtu for shale oil, 2,519 to 29,100 gal/MMBtu for ethanol fuel, and 14,000 to 75,000 gal/MMBtu for biodiesel from irrigated soy (Frac Attack 2010, 20).

Water Contamination

At first glance, one might expect the drilling and/or fracking of an oil or gas well might inevitably result in some contamination of groundwater in the area around the well. A variety of chemicals is typically added to the drilling fluid used in producing the well, and it would seem reasonable for some drilling line to release at least a minimal amount of those chemicals into groundwater. In fact, such a scenario should occur very rarely. The reason for this expectation is that the oil or gas well is drilled many hundreds or (much more commonly) thousands of feet into the underground rock, while the water table that contains essentially all groundwater in an area lies only a few hundred feet below the surface at the most (Explore Shale 2011). The loss of chemicals in a drilling fluid into the water table would, therefore, seem to be a very unlikely unoccurrence.

Yet complaints for residents in the area around oil and gas wells have long been common, arising even before the process of fracking was first used on such wells, but having become much more common since about 2005. These complaints focus on a number of troubling phenomena, the most dramatic of which may well be the case of "burning water." Some residents have reported that they have been able to set first to the water released from their household taps, presumably because of the methane present from natural gas trapped in the water. (To see a demonstration of this phenomenon, see Light Your Water on Fire from Gas Drilling, Fracking at http://www.youtube.com/watch?v=4LBjSXWQRV8.)

Although this complaint about fracking has received substantial attention and is certainly one of the most dramatic charges brought against the practice, it is unclear to what extent burning water and fracking are closely connected. A number of studies have shown, in fact, that flaming water is anything but an uncommon event in areas where oil and gas are produced. The apparent cause for the phenomenon is the presence of natural gas in shallow layers of earth around oil and gas wells, *whether they have been fracked or not*. Thus, while a legitimate complaint about fracking, it is not clear to what extent flaming water can be used as a major objection to the practice of fracking (Dunning 2011).

One question that might be asked about the contamination of water as a result of fracking is precisely what people mean by the term contamination. According to at least one study, most people are talking about the presence of methane in the water supply when they are talking about "contamination," and not about the dozens or hundreds of chemicals used in the fracking process (Fracking Caused Hundreds of Complaints about Contaminated Water in 4 States 2014). But both the systems by which states keep track of such complaints and the way in which complainants describe their experiences mask the precise meaning individuals ascribe to the term.

From a strictly scientific standpoint, the term *contamination* usually refers to the presence in water of substances normally not present in water and/or having the potential to cause harm to plants, animals (including humans), and other aspects of the environment. But reports of contamination by individuals and organizations often suggest a broader and somewhat more ambiguous experience. For example, the Pennsylvania Alliance for Clean Water and Air maintains a website listing specific complaints registered by individuals and communities that claim their water has been contaminated by fracking. As of late 2014, the website listed nearly 7,000 complaints, many of which came from groups of individuals, rather than single persons. Some of the experiences that complainants registered were the following: stomach pain; nausea; rashes on legs while showering; fatigue; mouth ulcers; sick and/or dead goats, dogs, cows, horses, and other farm animals; difficulty in breathing; elevated heart rate; lightheadedness; spitting up blood; farm and domestic animals refusing food and water; multiple myeloma; human mortality; air loss; increased sensitivity to chemicals; previously unidentified viruses; paralysis of legs; lesions on spinal cord; stress; shock; mental distress; and "constant fear of developing cancer" (Pennsylvania Alliance for Clean Water and Air 2014).

Defenders of fracking suggest that complaints such as those listed here must be assessed carefully. In the first place, many complaints have come from regions where conventional drilling has been practiced for many years. Any health problems that occur may be the result of these operations, rather than of fracking itself. Also, the wide range of problems that have been reported suggest that at least some individuals may be imagining or exaggerating health issues that are not really as severe as they may seem. The constant barrage of criticisms about fracking, supporters of the procedure say, may cause people to fantasize ailments that do not exist in reality. Overall, spokespersons for the oil and gas industry and their supporters sometimes simply deny that water contamination as a result of fracking even exists. In testimony before the U.S. House of Representatives Committee on Science, Space, and Technology, for example, Texas Railroad Commissioner David J. Porter stated categorically that "[t]hanks to the oil and gas industry's best practices and strict regulation and enforcement by the Railroad Commission, there has never been a confirmed case of groundwater contamination due to hydraulic fracturing in Texas" (Porter 2014). Since Texas maintains the most complete database of complaints about water contamination from fracking in the United States, Porter's comment is very significant in establishing the legitimacy of at least a vast majority of those complaints in the state (Begos 2014). Porter's view is apparently also born out by statements by representatives of a number of national and state governmental agencies and academic institutions, such as the U.S. Department of Energy, U.S. Geological Survey, U.S. Government Accountability Office, Environmental Protection Administration, Center for Rural Pennsylvania, Ground Water Protection Council, Massachusetts Institute of Technology, Stanford University, and Texas A&M University (Everley 2013c).

Intimately associated with the issue of water contamination as a result of fracking is the question as to what chemicals are used in the process. It is well known, in general, that oil and gas companies use a variety of different chemicals in order to increase the efficiency of the fracking process. These chemicals fall into a number of categories, such as acids (to dissolve rock), breakers (to reduce viscosity of a liquid), biocides (to prevent the growth of bacteria), buffers (to adjust for acidity of the well fluid), clay stabilizer (to prevent development of clayey structures), corrosion inhibitors (to prevent corrosion of metal components of the well), friction reducers, iron control (to prevent disruptive chemical changes caused by the presence of iron), and surfactants (to reduce fluid surface tension and improve flow of liquid in the well) (Hydraulic Fracturing 101 2014). At least 600 different chemicals are commercially available for achieving one or another of these objectives in the fracking process. (For one of the most complete lists of chemicals used in hydraulic fracturing, see Chemicals Used in Hydraulic Fracturing 2011, Appendix A.) The problem is that oil and gas companies have historically been reluctant to release to the general public (or even to regulatory agencies) the precise composition of chemicals used in any one field. The companies claim that such information is proprietary and can result in financial loss to the company if it becomes generally known.

Some companies, however, have begun to change their views and their policies about disclosing the names of the chemicals used in the fracking of their wells. They have come to realize that the success of their business depends to at least some extent on the trust that their consumers and the general public have in their products. If they can reduce public suspicions about the substances they are using in their operations, some companies believe, they may be able to avoid lawsuits, additional state and local regulations, and general bad feelings toward the company. One company that has led the way in providing full disclosure about their products is the Baker Hughes company, a supplier for the oil and gas drilling industry. In 2014, the company announced that it would make available on its website a complete listing of all chemicals used in its fracking products (Hydraulic Fracturing Chemical Disclosure Policy 2014; Energy Dept. Welcomes Fracking Chemical Disclosure 2014).

Detailed information about the composition of many fracked wells is now available on a website called FracFocus, maintained by the Ground Water Protection Council and Interstate Oil and Gas Compact Commission. The website is an interactive resource that allows one to enter information about a specific oil and/or gas well (such as the name of the state and county in which the well is located, the operator, and the API well number and name) for which information is desired. FracFocus then provides a complete list of all chemicals used at that well, along with CAS (Chemical Abstract Service) numbers for those chemicals. That information allows one to check detailed information about possible health and environmental hazards associated with each chemical included in the FracFocus list (see Find a Well 2014).

Once the chemicals used in a fracking operation are identified, a person can learn a great deal about the potential health threats posed by each chemical from a variety of sources. For example, TOXNET (Toxicology Data Network) is a service of the U.S. National Library of Medicine that provides access to 16 databases of hazardous chemicals. By simply entering the name of a chemical, one gains access to a host of information about its chemical and physical properties, along with its potential carcinogenic, mutagenic, teratogenic, and other health hazards to humans (TOXNET Databases 2014). Two other useful sources of information about the health effects of chemicals are the International Chemical Safety Cards

(ICSC), produced by the International Labour Organization (available at http://www.ilo.org/dyn/icsc/showcard.listCards2), and the NIOSH Pocket Guide to Chemical Hazards, published by the National Institute for Occupational Safety and Health (available at http://www.cdc.gov/niosh/docs/2005-149/pdfs/2005149.pdf).

Air Pollution

Critics of fracking often point to the risks posed to air quality by the procedure. They say that in some cases contamination of air can be as bad as, or even worse than, contamination of water, resulting in its own set of health issues for humans and other animals. The culprit most commonly named in contamination of the air is methane, the primary constituent of natural gas. Methane escapes from oil and gas wells at a number of points in the process of exploration, extraction, processing, and transportation. By far the greatest source of methane leaks at oil and gas wells is intentional venting of the gas to the atmosphere. Recall that oil and gas companies have traditionally allowed natural gas to escape from many wells simply because it is easier and less expensive to do so than to capture and store the gas. According to some studies, about 83 percent of all the methane that escapes from gas and oil production processes is released intentionally, with the remaining 17 percent escaping from leaks and accidental releases (Kiger 2014b).

But how serious is the problem of methane linkage from oil and gas wells? How much methane in actual amounts escapes from an oil or gas field during extraction? Surprisingly, little was known about the answer to that question until fairly recently. For many years, experts relied on a report published in 1996 by the Environmental Protection Agency (EPA) estimating that somewhere between 1 and 2 percent of all the methane produced in the United States escaped into the atmosphere (Methane Emissions from the Natural Gas Industry 1996, iv). By the second decade of the twenty-first century, however, a number of new studies began appearing suggesting that the EPA estimate was too low, perhaps much too low. A study of wells in the Denver–Julesburg Basin of the Rocky Mountains by researchers from the National Oceanic and Atmospheric Administration (NOAA) and the University of Colorado, found that about 4 percent of the methane produced from oil and gas wells was escaping into the atmosphere, about twice that of the 1996 EPA report (Pétron et al. 2012; Tollefson 2012).

In the years following the NOAA–Colorado study, additional research found even more striking data about the loss of methane from oil and gas wells. These data showed loss rates of as high as 14 and 17 percent in some cases, approaching 10 times the 1996 EPA estimates (Report: Methane Emissions at "High Levels" in Uintah Basin 2013). Then, in late 2013, researchers at the University of California at San Diego published their review of data from a variety of sources about methane emissions in the United States. They concluded that those emissions could be anywhere from 2.3 to 7.5 times greater than the "official" estimates of the highly respected Emissions Database for Global Atmospheric Research (Miller et al. 2013; also see Brandt et al. 2014).

Data suggesting high levels of methane linkage are not accepted uncritically by all observers. Representatives of the oil and gas industry in particular tend to find errors in the conduct of studies leading to these conclusions and/or are able to cite other research that shows very different trends. For example, a writer for the Internet site Energy in Depth (created by the Independent Petroleum Association of America) has called attention to studies that found methane linkages in the range of 1.5 percent "or less than that," comparable to the 1996 EPA study. The writer concludes his review of the study with the observation that:

> The activist fear-mongering about methane emissions has been exposed as fraudulent by the most comprehensive research on the subject to date, including data that

> incorporates the first-ever direct measurements of methane emissions from various segments of the production process. Perhaps now we can all come together, take a deep breath, and recognize the clear economic and environmental benefits of natural gas from shale. (Everley 2013a)

One of the interesting trends in research on air contamination resulting from fracking operations has been the increasing evidence of air pollutants other than methane itself as potential sources of human health problems. In 2012, for example, researchers at The Endocrine Disruption Exchange in Paonia, Colorado, reported on their study of compounds known collectively as *non-methane hydrocarbons* (NMHCs) in the effluent from fracked oil and gas wells. NMHCs are hydrocarbons that occur in natural gas at a much lower concentration than does methane itself: ethane, propane, butane, ethene (ethylene), propene (propylene), and ethyne (acetylene), for example. Little attention was paid to the presence, concentration, and health effects of these compounds until 2012. But they appear to pose their own significant and identifiable human health issues. As just one example, researchers found methylene chloride, which causes damage to the nervous system, present almost three quarters of the time in their sampling. They concluded that NMHCs "should be examined further given that the natural gas industry is now operating in close proximity to human residences and public lands" (Colborn et al. 2012; Song 2012).

Many of the chemical compounds found in polluted air in general are also found in areas where fracking is occurring. These compounds include sulfur dioxide, oxides of nitrogen, carbon monoxide, polyaromatic hydrocarbons (PAHs), volatile organic compounds (VOCs), and PM10 and PM2.5 (particulate matter less than 10 and 2.5 micrometers in diameter, respectively). In its annual National Emissions Inventory for 2012, for example, the EPA reported that the oil and gas industry was responsible for the release of more than 1.688 billion tons of VOCs during production, more than seven times as much as the next largest polluter, the storage and transfer industry, and more than a thousand times that of the rest of the mining industry (Weinhold 2012, A276). VOCs have been associated with a wide variety of medical conditions (at least partly because the category includes so many different organic compounds), such as eye, nose, and throat irritation; headaches; loss of coordination; nausea; damage to liver, kidney, and central nervous system; conjunctival irritation; nose and throat discomfort; headache; allergic skin reaction; dyspnea; declines in serum cholinesterase levels; fatigue; and dizziness. Some components of VOCs have also been implicated in the development of cancers in experimental animals.

Finely divided particulate matter (PM) has also been shown to have a number of health problems. Such pollutants have a tendency to become lodged in the respiratory system, causing health concerns such as decreased lung function, in general; irregular heartbeat; nonfatal heart attacks; premature death in people with heart or lung disease; and aggravated asthma (Particulate Matter 2014).

One recent study by the Center for Public Integrity, InsideClimate News, and The Weather Channel, produced a number of troubling findings about air pollution resulting from fracking in the Eagle Ford Play of Texas. Among those findings were the following:

- The state of Texas maintains only five permanent air monitoring systems in the 20,000-square-mile Eagle Ford Play, a number clearly inadequate to obtain the necessary raw data about possible air contamination in the region.
- The state allows oil and gas companies to self-monitor their operations, resulting in the fact that it does not even know how many fracking operations are being conducted in the states, where they are located, or what environmental effects they are having on residents of the region.

- Companies that do violate state regulations are rarely fined, with only two penalties having been assessed in the 164 documented violations found by the investigating team in the period of the study. The largest fine assessed for the violations was $14,250.
- The Texas legislature reduced the budget for the Texas Commission on Environmental Quality (TCEQ) by a third during its 2014 session.
- The release of toxic air pollutants increased by 100 percent between 2009 and 2014.
- Air quality monitoring systems are unequally distributed in Texas with urban areas such as Dallas receiving many more such systems per resident than areas close to fracking operations, such as those in the Eagle Ford Play (Morris, Song, and Hasemyer 2014).

Fracking supporters have questioned the complaints about air pollution from fracking, just as they have complaints about water contamination. On its website, for example, the "nonprofit research organization" The Heartland Institute, claims that concerns about air pollution caused by fracking are "exaggerated." In support of this claim, the organization quotes Bryan Shaw, chair of the TCEQ, on the subject. "After several months of operation," Shaw is quoted as saying, "state-of-the-art, 24-hour air monitors in the Barnett Shale area are showing no levels of concern for any chemicals. This reinforces our conclusion that there are no immediate health concerns from air quality in the area, and that when they are properly managed and maintained, oil and gas operations do not cause harmful excess air emissions" (Smith 2012; for Shaw's original statement, see A Commitment to Air Quality in the Barnett Shale 2010).

Earthquakes

The question as to the relationship between the extraction of gas and oil and seismic events (earthquakes), if any, has interested researchers for at least the last half century. During the 1960s, that question arose in a somewhat different context, the use of injection wells for the disposal of radioactive wastes. At the time, scientists were exploring systems for burying or otherwise disposing of the waste materials produced by research on atomic weapons and at nuclear power plants. One proposal was to force those (liquid) wastes into deep wells that had previously been used for other purposes, such as the collection of water for human use. Soon after this method of disposal was begun, the number of seismic events (earthquakes) in the region surrounding the injection wells begin to increase significantly. Researchers came to the conclusion that the liquid wastes were acting as a lubricant on rocks deep underground, causing them to slip and fracture, thus producing the earthquakes (see, for example, Evans 1966). (The disposal of radioactive wastes in wells was discontinued shortly after this study was conducted, but debate continues today as to the viability of the procedure for the disposal of such wastes. See Underground Injection Control Program 2013.)

The effect of oil and gas drilling operations on earth movements has also been a matter of some interest to researchers, although much more so in the past decade than in previous history. From a simple-minded point of view, one would expect at least some effects of drilling on seismic events since oil and gas are often found along unstable rock formations deep underground, and the act of drilling itself might be expected to cause some shifts in the rock. In fact, research has shown that conventional oil and gas wells appear to have little or no effect on observable instability of rock formations in surrounding regions. Some evidence suggests that hydraulic fracturing *may* be the cause of some seismic events, but the data for such conclusions is typically uncertain and ambiguous. For example, one study of a series of about 50 earthquakes in Garvin County, Oklahoma, in January 2011, was investigated by researchers at the Oklahoma Geological Survey to determine the involvement, if any, of fracking operations in the area.

Those researchers concluded that "the uncertainties in the data make it impossible to say with a high degree of certainty whether or not these earthquakes were triggered by natural means or by the nearby hydraulic fracturing operation" (Holland 2011).

Evidence for the involvement of injection wells used during fracking operations with seismic events is much more compelling. Recall that very large quantities of water are used in fracking operations and that one common method of disposing of the wastewater produced during fracking is emptying into deep wells. The issue of the relationship between fracking injection wells and seismic events is, thus, similar to that of radioactive waste disposal and seismic events.

For some observers, especially those associated with the oil and gas industry, the connection between earthquakes and deep well injection of fracking wastes is unproved or questionable, at best. In one commentary on this issue, an anonymous writer points out that there are more than 144,000 injection wells in the United States that have been used safely by a variety of industries for decades. The writer quotes the EPA as saying that injection wells are "a safe and inexpensive option for the disposal of unwanted wastes," and that the reports of individual seismic events associated with the use of injection wells by the oil and gas industry are few and far between. The writer suggests that the issue is not one that people should worry about because the "minute vibrations" that occur during such events "may not even be detectable to humans" (Does Fracking Cause Earthquakes? 2014).

The preponderance of scientific evidence at this point in time, however, seems to point to some connection between the use of injection wells for the disposal of fracking wastewater and seismic events. This evidence has been coming in from oil and gas extraction sites across the United States. One of the most extensively studied series of seismic events has been those taking place in central Oklahoma beginning in January 2009. Between that date and the end of 2013, more than 200 seismic events of intensity greater than 3.0 on the Richter scale were recorded, including the largest earthquake ever recorded in the state, which struck near Prague, Oklahoma, on November 5, 2011. That quake registered 5.6 on the Richter scale. The increase in seismic events was striking because the state had experienced an average of fewer than three 3.0 magnitude earthquakes between 1975 and 2008, a number that rose to 40 per year in the period between 2009 and 2013 (Koontz 2013).

A similar pattern was observed in another state without a history of earthquakes: Ohio. Between January 2011 and February 2012, more than 100 seismic events with magnitude between 0.4 and 3.9 were recorded in the Youngstown area, a region with no known history of earthquakes. According to a report by geologist Won-Young Kim, of Columbia University, the probable explanation for the seismic events was the disposal of wastewater into injection wells along a fault line underlying the area (Kim 2013).

As of late 2014, the U.S. Geological Survey was still unwilling to express a definite determination between the injection of wastewater from fracking operations into wells and seismic events. Their position at the time was that, based on evidence that had been obtained thus far, further research was needed about this relationship (Induced Earthquakes 2014; see also Ellsworth 2013)

As of 2014, the potential hazards posed by injection well disposal of fracking wastewater appear to be increasingly significant. At the annual meeting of the Seismological Society of America, a number of papers were presented that suggested that seismic events associated with deep well disposal were more common, more severe, and more difficult to predict than researchers had previously thought. Some evidence was also presented suggesting that the range of such seismic events from injection wells can be considerably greater than previously listed (Kiger 2014a; Wastewater Disposal May Trigger Quakes at a Greater Distance than Previously Thought 2014;

for a good general discussion of this issue, see Connelly, Barer, and Skorobogatov 2014; Man-Made Earthquakes Update 2014).

The debate over fracking wastewater disposal and seismic events is likely to continue, but at least some governmental agencies have decided to act on the information that is currently available. For example, the state of Colorado decided in June 2014 to issue a temporary ban on the disposal of fracking wastewater at one well near which a number of seismic events had been recorded (COGCC Halts Activity at Injection Well 2014; a summary on restrictions on the use of injection wells for the disposal of fracking wastewater as of mid-2014 can be found at Rushton and Castaneda 2014).

Aesthetic and Related Disturbances

Another large category of objections to fracking operations can be classified as *nuisance* complaints. Legally, the term *nuisance* refers to any action that is harmful or annoying to an individual person or discrete group of individuals (a *private nuisance*) or to a larger community of individuals (a *public nuisance*) (Nuisance 2014). Consider some of the actions that are commonly associated with a fracking operation at an oil or gas field. In many cases, roads may have to be built to the fracking site. Land may have to be cleared and private property may have to be condemned by the state or county to gain rights of drilling and access to land. In these processes, aesthetically appealing areas on which people built homes for their solitude, views, access to nature, or other purposes may be destroyed.

Traffic in a region where fracking is occurring may also increase significantly. Truck traffic may increase in order to bring in the equipment and material used in fracking and to remove the wastewater produced by the operation. This truck traffic may necessitate expensive repairs of city and county roads, an expense that must usually be paid by residents, not the fracking company. These harms to the land may cause property values to decrease and can increase the difficulties in obtaining normal homeowners insurance and reasonable mortgage rates. The presence of fracking operations may also have an effect on the economy of a region, such as discouraging the creation of new businesses, driving away customers for existing businesses, reduce the appeal of recreational opportunities, and result in deferred maintenance of existing structures in the region. (Some of the best descriptions of nuisance complaints because of fracking can be found on the websites of local groups that have been organized to oppose such operations in their neighborhoods. See, for example, DeBerry 2013; Gleeson 2011; Woodrum 2014; Woods 2014.)

Laws and Regulations

Many of the most important federal laws were passed to protect human health and the natural environment from pollution almost a half century ago. These legacy laws included the National Environmental Policy Act of 1970, Clean Air Act of 1970, Occupational Safety and Health Act of 1970, Clean Water Act of 1972, Safe Drinking Water Act of 1974, Federal Land Policy and Management Act of 1976, Resource Conservation and Recovery Act of 1976, Surface Mining Control and Reclamation Act of 1977, and Comprehensive Environmental Response, Compensation and Liability Act of 1980. All of these acts have been updated and amended a number of times since their original adoption (Environmental Law 2014). None of these laws mentions potential health or environmental risks associated with fracking or related oil and gas operations. That omission is easy to understand, since fracking did not become a significant industrial activity until after the turn of the twenty-first century.

In fact, the EPA on at least one occasion specifically explained, when confronted with the issue, that fracking was generally regarded as a safe operation that used chemicals that posed no threat to human health or to the natural environment.

When asked by the Legal Environmental Assistance Foundation (LEAF) to take action on the regulation of fracking operations at an Alabama oil site in 1995, EPA administrator Carol Browner wrote to the organization that "EPA does not regulate—and does not believe it is legally required to regulate—the hydraulic fracturing of methane gas production wells. . . . There is no evidence that the hydraulic fracturing at issue has resulted in any contamination or endangerment of underground sources of drinking water." (See the original letter at http://energyindepth.org/docs/pdf/Browner-Letter-Full-Response.pdf. Accessed on July 3, 2014.)

In fact, the exchange between LEAF and EPA instigated a landmark legal case about the regulation of fracking when the former organization appealed Browner's decision to the U.S. Court of Appeals for the Eleventh Circuit. That court ruled, in fact, that EPA was not only authorized to regulate fracking, but in fact was required to do so (*Legal Environmental Assistance Foundation, Inc., Petitioner,* vs. *United States Environmental Protection Agency, Respondent* 1997). The court's ruling represented something of a Pyrrhic victory for LEAF since the rules established by the Alabama State Oil and Gas Board for fracturing were eventually approved by the EPA even though LEAF felt that they fell far short of providing the protection needed for fracking operations.

Not much changed with regard to laws and regulation of fracking for almost a decade, to a considerable extent because of a comprehensive study by the EPA that found that fracking posed no harm to water quality. In a summary of that report, the EPA stated that

> EPA found no confirmed cases that are linked to fracturing fluid injection into CBM [coal-bed methane] wells or subsequent underground movement of fracturing fluids. Further, although thousands of CBM wells are fractured annually, EPA did not find confirmed evidence that drinking water wells have been contaminated

> by hydraulic fracturing fluid injection into CBM wells. (Evaluation of Impacts to Underground Sources of Drinking Water 2004)

The EPA further concluded that, as a result of its study, "additional or further study is not warranted at this time."

The methodology and conclusions of the EPA study were questioned both at the time the study was completed and the years since. One EPA researcher, Weston Wilson, for example, wrote at the time that

> EPA's conclusions are unsupportable. EPA has conducted limited research reaching the unsupported conclusion that this industry practice needs no further study at this time. EPA decisions were supported by a Peer Review Panel; however five of the seven members of this panel appear to have conflicts-of-interest and may benefit from EPA's decision not to conduct further investigation or impose regulatory conditions. (See Wilson's original letter at http://media.trb.com/media/acrobat/2004-10/14647025.pdf. Accessed on July 3, 2014.)

The first opportunity the U.S. Congress had to enact legislation dealing with fracking operations following the EPA report came a year later when it considered legislation that was to become the Energy Policy Act of 2005, a wide-ranging act that, as the title suggests, dealt with a diverse number of issues related to the production and consumption of energy in the United States. One of the 1840 sections of the act dealt with hydraulic fracturing, a section that essentially *excluded* hydraulic fracturing operations from the provisions of the Safe Drinking Water Act of 1974. The section later became popularly known as the *Halliburton loophole* because it apparently gave free-rein to the Halliburton company, one of the largest providers of fracking operations in the world. At the time, the nation's vice president was Richard Cheney, who had spent most of his professional life as a high-ranking employee of the Halliburton corporation. Some critics suggested that Cheney may have worked to include Section 322 of the Energy Policy Act to protect an important activity of his former company. The so-called loophole remains in effect as of late 2014 (Counts 2013; The Halliburton Loophole 2014).

In 2009, U.S. Senator Bob Casey (D-PA), and U.S. Representatives Diana DeGette (D-CO), Maurice Hinchey (D-NY), and Jared Polis (D-CO) jointly introduced legislation in the United States to remove the Halliburton loophole from the Energy Policy Act of 2005. Their bill, officially called the Fracturing Responsibility and Awareness of Chemicals Act, became popularly known as the FRAC Act of 2009. The bill failed to receive consideration in the 111th Congress, but was reintroduced in the 112th and 113th sessions of the Congress, where it also died without being considered. Thus, as of late 2014, no federal directive exists for the regulation of fracking in the United States at the federal level. In fact, the practice is specifically excluded from regulation not only in the Energy Policy Act of 2005, but also in amendments to a variety of other environmental legislation, including the Safe Drinking Water Act; the Resource Conservation and Recovery Act; the Emergency Planning and Community Right-to-Know Act; the Clean Water Act; the Clean Air Act; the Comprehensive Environmental Response, Compensation, and Liability Act; and the National Environmental Policy Act (Brady and Crannell 2012, 43–52).

The absence of federal laws and regulation does not mean that hydraulic fracturing operations are not subject to such monitoring by other governmental levels. Indeed, a number of states and local municipalities have now adopted a variety of laws and regulations relevant to fracking operations. Tracking these laws and regulations is a challenge because attitudes are changing so rapidly that the list of states and communities with hydraulic fracturing laws and regulations changes almost monthly. The following review is, therefore, at least partially obsolete even before it appears in print. The interested reader should consult a search engine for the most recent list and description of such legislation.

State and local laws and regulations generally refer to one of four major steps in the oil and gas extraction process: predrilling, groundwater and surface water impact, liquid wastes and fluids, and solid wastes (Fracking Regulations by State 2014). For example, the state of Colorado requires that companies test for and report on a number of site characteristics before drilling, such as pH and alkalinity and the concentration (if any) of chemicals such as chloride, fluoride, sulfate, and heavy metals (such as arsenic, cadmium, and iron). This information provides a baseline against which later measurements can be compared to see what changes, if any, fracking has made in those site characteristics.

Probably the greatest current concern expressed in state laws is the character of wastewater and the methods by which it is disposed. Companies may be required to report to state regulatory agencies on the presence of potentially hazardous chemicals such as benzene, toluene, xylene, ethylene glycol, phenols, a great variety of metals, certain anions (such as chloride and sulfate), and radioactive isotopes such as radium-226 and radium-228. Some of the chemicals for which acceptable limits are listed by the state of Colorado, as an example, are benzene, toluene, ethylbenzene, xylenes, acenaphthene, anthracene, benzo(A)anthracene, benzo(B)fluoranthene, benzo(K)fluoranthene, benzo(A)pyrene, chrysene, dibenzo(A,H)anthracene, fluoranthene, fluorene, indeno(1,2,3,C,D)pyrene, napthalene, pyrene, chlorides, sulfates, arsenic, barium, boron, cadmium, chromium(III), chromium(VI), copper, lead, mercury, nickel, selenium, silver, and zinc. (E&P Waste Management 2014; A complete interactive list of state regulations is available at FracFocus, http://fracfocus.org/regulations-state. Accessed on July 3, 2014.)

One trend that seems to be occurring in the adoption of state fracking laws is a more rigorous stance on the types of testing and monitoring that must be done to gain a permit to use hydraulic fracturing. In contrast to the somewhat more *laissez-faire* attitude of the federal government, states appear to have adopted the position that oil and gas companies should be much more responsible for the health and environmental consequences of their operations than may have been true in the past. For example, when the state of Illinois adopted new fracking legislation in 2013, some observers called it the "toughest fracking law in the nation" because of its requirements that companies provide a complete description of all chemicals used in the fracking process and submit detailed before-and-after reports of environmental conditions around the well (Babwin 2013). All of which was not to say that stakeholders were universally delighted with the bill. As a member of the group Southern Illinoisans Against Fracturing Our Environment said after the bill passed, "[t]his bill was written by industry and parties that have a vested interest. . . . We have no say in our own water. . . . We are totally helpless" (Lester 2013).

A number of local communities in the United States have also taken action with regard to fracking operations. These actions range from the enactment of outright bans on hydraulic fracturing within the boundaries of the governmental unit to certain types of regulations and restrictions on fracking to expressions of support for bans, moratoria, or regulations on the practice. As of late 2014, the number of governmental entities in the United States that have taken such actions number 418, according to the Keep Tap Water Safe website. (This number obviously changes over time. The state with by far the largest number of governmental entities having adopted some form of regulation is New York, where more than 200 towns, villages, cities, counties, and other jurisdictions have enacted some form of action on fracking operations (Keep Tap Water Safe 2014).

Some examples of the actions taken by local communities include the following:

- The city council of Canandaigua, New York, voted in June 2014 to permanently ban the exploration or drilling for or transfer, storage, treatment, or disposal of natural gas within the city boundaries (Ordinance #2014-006 2014).
- In May 2014, Santa Cruz County became the first county in California to ban all exploration for and extraction of oil and gas within the county (Board of Supervisors Resolution Amending the Santa Cruz County General Plan Regarding Prohibition on Oil and Gas Exploration and Development 2014).
- Mora County, New Mexico, one of the smallest (5,000 inhabitants) and poorest counties in the state, became the first county in the United States in April 2013 to ban all exploration for and extraction of oil and gas in the county (Ordinance 2013-01 2013).
- The town of Penn Yan, New York, adopted a restricted action against fracking, declaring that it would not accept fracking wastes at its local wastewater treatment plant (Untitled document [Penn Yan, New York] 2014).
- The borough council of West Homestead, Pennsylvania, approached the fracking question in a somewhat different way than an outright ban or moratorium, a method used by a number of other municipalities in the United States. It adopted a bill of rights for residents of the area that includes a Right to Water, Rights of Natural Communities, Right to a Sustainable Energy Future, Right to Self-Government, Rights as Self-Executing, and a statement of People as Sovereign. The council then noted that these rights de facto prohibited the practice of exploration for or extraction of oil and gas within borough boundaries (Ordinance #659 2011).
- A number of municipalities have taken what is probably the least aggressive approach to dealing with hydraulic fracturing issues, namely requesting that their representative or senator to the U.S. Congress or some other federal body consider or

> take action to prevent fracking operations or deep well injection disposal from occurring within their governmental boundaries. See for example the resolution adopted by the council of Meyers Lake Village, Ohio, requesting the U.S. Department of Labor to take such actions and the letter from the Anna (Illinois) city council asking the speaker of the Illinois House of Representatives to aid in the advancement of an anti-fracking bill in the state senate. (Village Council Resolution No. 2012-614 2012; High-Volume Horizontal Hydraulic Fracturing 2012)

The legal maneuvering over hydraulic fracturing has raised a fundamental question for legislatures and the courts: Who gets to decide the conditions under which fracking should be allowed to take place in an oil or gas field and who gets to decide whether or not hydraulic fracturing and deep well injection disposal should even be permitted? The complex issues surrounding these questions are just barely getting resolved in 2014 since fracking itself is still in its earliest stages and the federal government, states governments, tribal councils, and local municipalities have yet to decide who has the authority to regulate each aspect of oil and gas exploration, drilling, transportation, waste disposal, and other activities related to the industry.

Perhaps the earliest legal debate over the right of the state or local government to regulate oil and gas drilling came in the early 1990s in the case of *La Plata County vs. Bowen/Edwards, Associates, Inc.* The case arose when the board of county commissioners for La Plata County passed an ordinance requiring oil and gas companies to get a special land use permit to conduct drilling operations in the county. The Bowen/Edwards drilling company sued to have that requirement overturned, arguing that the Colorado Oil and Gas Conservation Commission had sole authority to require drilling permits in the state. After a series of decisions for and against the Bowen/Edwards position, the Colorado Supreme Court eventually ruled in favor of the La Plata County position, stating that there was no justification for the state to interfere in what was essentially a local question about land use in the county.

The Colorado case set a precedent for similar cases that were to arise in other states over the next decade or more. One of the most important of those cases arose in the state of Pennsylvania, where hydraulic fracturing had become a widespread practice. Fracking operations had become popular and well received in a number of communities that saw the potential economic benefits offered by the practice, in spite of any health and environmental concerns that might be associated with the process. But they had become anathema among a number of cities, towns, counties, and other municipalities whose judgment was that those same health and environmental issues were more significant than any possible economic gains. By 2013, a number of municipalities in the latter category had adopted ordinances or regulations that placed a moratorium on fracking, instituted specific regulations about the practice, or simply banned fracking from the city, town, or county.

As in most other states, the legal question arose as to whether the state as a whole or individual communities had the ultimate authority to ban fracking or not. The conflict was made even more clear in Pennsylvania in 2012 when the state legislature passed Act 13, a bill designed to set out state policy on the use of fracking operations by the oil and gas industry. A major provision of the act was that local communities were prohibited from passing bans or other restrictions on fracking operations; they were, in effect, proscribed from enforcing local zoning laws that might be used to interfere with fracking operations on lands within their jurisdiction.

Environmentalists and a number of other interested groups were outraged at this action, arguing that Act 13 prevented the citizens of Pennsylvania from "making critical decisions for themselves . . . from protecting the health, safety, and welfare of their residents in the face of unrelenting corporate assaults" (PA Legislature Pre-empts Communities on Fracking

2012). Seven Pennsylvania communities felt so strongly about the matter that they brought suit against the state in a case that was to become known as *Robinson Township, et al. vs. Commonwealth of Massachusetts* (J-127A-D-2012 2013). In December 2013, the Supreme Court of Pennsylvania decided that portions of Act 13 were unconstitutional because they violated the Environmental Rights Amendment of the state constitution. The sections that the court invalidated included restrictions such as:

- The department shall waive the distance restrictions [as to the safe and appropriate siting of a well for the protection of the environment] upon submission of a plan identifying additional measures, facilities or practices to be employed during well site construction, drilling and operations necessary to protect the waters of this Commonwealth. §3215 (b) (4)
- Notwithstanding any other law to the contrary, environmental acts are of Statewide concern and, to the extent that they regulate oil and gas operations, occupy the entire field of regulation, to the exclusion of all local ordinances. The Commonwealth by this section, preempts and supersedes the local regulation of oil and gas operations regulated by the environmental acts, as provided in this chapter. §3303
- In addition to the restrictions contained in sections 3302 (relating to oil and gas operations regulated pursuant to Chapter 32) and 3303 (relating to oil and gas operations regulated by environmental acts), all local ordinances regulating oil and gas operations shall allow for the reasonable development of oil and gas resources. §3304 (J-127A-D-2012 2013, 161; for reactions to the court's decision on this matter, see Raichel 2013 and related blogs.)

A second important court decision was rendered in June 2014 by New York state's highest court, the Court of

Appeals. The decision involved two appeals in cases that had been working their way up through the New York court systems since 2011. In one case, the Norse Energy company had been acquiring land in the town of Dryden since 2006 with the intent of eventually drilling for oil and gas and, presumably, using fracking operations in the process. In the second case, the Cooperstown Holstein Corporation (CHC) had signed two leases with landowners for a similar purpose of exploring for oil and gas on property owned by those landowners. When both the town of Dryden and the town of Middlefield, where those leases had been executed, decided in 2011 to pass ordinances banning hydraulic fracturing, Norse and CHC both filed suit against the respective towns in which their leases were located.

These cases differed somewhat from the Pennsylvania case cited earlier because the state of New York had not yet made a final policy decision about fracking. A moratorium had been in place since 2008 while the state Department of Health conducted a review of possible health effects of fracking. Before that review could be completed, the Court of Appeals ruled in favor of the defendants in both cases decided in June 2014. The court said that local communities had the right to use zoning laws to ban exploration for and/or extraction of oil and natural gas on lands within their boundaries (Taylor and Kaplan 2014; for the court's ruling, see http://www.nycourts.gov/ctapps/Decisions/2014/Jun14/130-131opn14-Decision.pdf 2013).

The New York ruling is also expected to have more wide-reaching effects than will the Pennsylvania ruling. In the later case, hydraulic fracturing had (and has) been widely used throughout the state where it has provided one of the most extensive accesses to the rich Marcellus Play. In New York state, by contrast, little exploration or extraction had yet occurred because the state itself had not made any decision about its position on fracking, and more than 200 hundred towns, villages, cities, and other jurisdictions had already decided to ban the practice or, at the very least, to hold off on giving its approval. As a result, some observers expressed the view that the June 2014 court decision would essentially exclude oil and gas activities from the state of New York for the foreseeable future (Dawson 2014). At least some energy vowed to fight back in the state legislature against the court ruling, however. Norse Energy's attorney in the case, for example, warned that the court's decision would have a "very chilling effect" on research and that it would be "very hard for operators to justify spending hundreds of millions of dollars to come in and not have regulatory certainty." An attorney for CHC went on to ask "We are going to let 932 towns decide the energy policy of New York state" (Mufson 2014).

Conclusion

The oil and gas process known as hydraulic fracturing has been known for more than a century. But its widespread use in the industry is no more than about a decade old. During that time, fracturing has been responsible for the explosive growth in oil and gas production from shale and other rock formations, bringing enormous profits to the industry, a dramatic increase in jobs in the United States, and the potential for energy independence for this country. But the process is also responsible for a number of potentially harmful effects on human health and the natural environment. Concerns about these effects are serious enough that a number of states, cities, and other governmental bodies have banned the use of fracking within their territories or, at the least, imposed significant restrictions on the conditions under which fracking can be used. But it is still early days for fracking in the oil and gas industry, and no one can yet say with any certainty how this technologically important process will contribute to the future energy equation of the United States, other nations, and, indeed, the world. Nor is it yet possible to predict the ways in which the troubling aspects of hydraulic fracturing will be handled by American society or by other countries of the world.

References

Anderson, Richard. "How American Energy Independence Could Change the World," BBC News Business, http://www.bbc.com/news/business-23151813. Accessed on June 24, 2014.

"As Gas Prices Pinch, Support for Oil and Gas Production Grows." 2012. Pew Research Center for the People & the Press, March 19. http://www.people-press.org/2012/03/19/as-gas-prices-pinch-support-for-oil-and-gas-production-grows/. Accessed on June 26, 2014.

"Australia's Victoria State Bans Coal Seam Gas 'Fracking'." Reuters, http://www.reuters.com/article/2012/08/24/us-australia-csg-victoria-idUSBRE87N06D20120824. Accessed on June 24, 2014.

Babwin, Don. 2013. "Illinois Gas Drilling Rules: Governor Pat Quinn Signs New Fracking Regulations into Law," *Huffington* Post Green, June 17. http://www.huffingtonpost.com/2013/06/17/illinois-gas-drilling-rules-fracking_n_3455668.html. Accessed on July 3, 2014.

Bailey, David. "In North Dakota, Hard to Tell an Oil Millionaire from Regular Joe," Reuters, http://www.reuters.com/article/2012/10/03/us-usa-northdakota-millionaires-idUSBRE8921AF20121003. Accessed on June 25, 2014.

Begos, Kevin. 2014. "4 States Confirm Water Pollution from Drilling." USA Today, January 5, http://www.usatoday.com/story/money/business/2014/01/05/some-states-confirm-water-pollution-from-drilling/4328859/. Accessed on June 28, 2014.

"Benefitting from Unconventional Oil." 2012. Bill Lane Center for the American West, Stanford University, April.

http://web.stanford.edu/group/ruralwest/cgi-bin/projects/headwaters/Bakken-Energy-Report-Headwaters-BLC-120424.pdf. Accessed on June 25, 2014.

"Board of County Commissioners." 1992. La Plata County, Colorado, Petitioner, *vs. Bowen/Edwards Associates, Inc., Respondent*, No. 90sc516. Supreme Court of Colorado, June , 1992. LexisNexis, http://legisource.net/wp-content/uploads/2012/09/830_P_2d_1045__1055__Colo__1.pdf. Accessed on July 4, 2014.

"Board of Supervisors Resolution Amending the Santa Cruz County General Plan Regarding Prohibition on Oil and Gas Exploration and Development." April 30, 2014. http://documents.foodandwaterwatch.org/doc/Frack_Actions_SantaCruzCountyCA.pdf#_ga=1.41040033.229352318.1403631732. Accessed on July 4, 2014.

"BP Energy Outlook 2030." January 2013. http://www.bp.com/content/dam/bp/pdf/statistical-review/EnergyOutlook2030/BP_Energy_Outlook_2030_Booklet_2013.pdf. Accessed on June 24, 2014.

Brady, William J., and James P. Crannell. 2012. "Hydraulic Fracturing Regulation in the United States: The Laissez-Faire Approach of the Federal Government and Varying State Regulations," *Vermont Journal of Environmental Law* 14 (1): 39–70.

Brandt, A. R., et al. 2014. "Methane Leaks from North American Natural Gas Systems," *Science,* 343 (6172): 733–735.

Brown, Erica, et al. 2013. "Public Opinion on Fracking: Perspectives from Michigan and Pennsylvania," The Center for Local, State, and Urban Policy, University of Michigan, May. http://closup.umich.edu/files/nsee-fracking-fall-2012.pdf. Accessed on June 26, 2014.

Brown, Travis H. 2014. "Fracking Fuels and Economic Boom in North Dakota," Forbes, January 29. http://www.forbes.com/

sites/travisbrown/2014/01/29/fracking-fuels-an-economic-boom-in-north-dakota/. Accessed on June 25, 2014.

Bryce, Robert. 2008. *Gusher of Lies: The Dangerous Delusions of "Energy Independence."* New York: Public Affairs.

Burnett, H. Sterling. "How Fracking Helps Meet America's Energy Needs," National Center for Policy Analysis, http://www.ncpa.org/pub/ib132. Accessed on June 22, 2014.

Carlyle, Ryan. "How Big Are the Currently Known Oil Reserves and What Are the Chances of Finding New Ones?" Forbes, http://www.forbes.com/sites/quora/2013/03/27/how-big-are-the-currently-known-oil-reserves-and-what-are-the-chances-of-finding-new-ones/. Accessed on June 23, 2014.

"Chemicals Used in Hydraulic Fracturing." 2011. United States House of Representatives Committee on Energy and Commerce, April.http://democrats.energycommerce.house.gov/sites/default/files/documents/Hydraulic-Fracturing-Chemicals-2011-4-18.pdf. Accessed on June 28, 2014.

"COGCC Halts Activity at Injection Well: Seeks Additional Review." 2014. Oil and Gas Conservation Commission, June 24. http://dnr.state.co.us/newsapp/press.asp?pressid=9008. Accessed on July 2, 2014.

Colborn, Theo, et al. 2012. "An Exploratory Study of Air Quality near Natural Gas Operations." *Human and Ecological Risk Assessment* 20 (1): 86–105.

"A Commitment to Air Quality in the Barnett Shale." 2010. Texas Commission on Environmental Quality, Fall. http://www.tceq.texas.gov/publications/pd/020/10-04/a-commitment-to-air-quality-in-the-barnett-shale. Accessed on June 30, 2014.

Connelly, Kelly, David Barer, and Yana Skorobogatov. 2014. "How Oil and Gas Disposal Wells Can Cause Earthquakes." StateImpact, June 19. http://stateimpact.npr.org/texas/tag/earthquake/. Accessed on July 2, 2014.

Considine, Timothy J., Robert W. Watson, and Nicholas B. Considine. 2011. "The Economic Opportunities of Shale Energy Development," Manhattan Institute for Policy Research, http://www.manhattan-institute.org/html/eper_09.htm. Accessed on June 25, 2014.

Counts, Nicole. 2013. "Should the 'Halliburton Loophole' Be Revoked from the Energy Policy Act of 2005?" Law Street. October 5. http://lawstreetmedia.com/issues/energy-and-environment/should-the-halliburton-loophole-be-revoked-from-the-energy-policy-act-of-2005/. Accessed on July 3, 2014.

Dawson, Evan. 2014. "Connections: The Future of Fracking in New York." WXXI News, July 3. http://wxxinews.org/post/connections-future-fracking-new-york. Accessed on July 5, 2014.

DeBerry, Candy. 2013. "Drilling, Fracking Not Safe." Observer-Reporter.com, January 31. http://www.observer-reporter.com/article/20130131/OPINION02/1301 39889#.U7RAwfldV8E. Accessed on July 2, 2014.

"Does Fracking Cause Earthquakes?" Energy Answered, http://energyanswered.org/questions/does-fracking-cause-earthquakes. Accessed on July 1, 2014.

Dunning, Brian. 2011. "All about Fracking." Skeptoid, September 13. http://skeptoid.com/episodes/4275. Accessed on June 27, 2014.

"E&P Waste Management." State of Colorado, http://cogcc.state.co.us/RR_Docs_new/rules/900Series.pdf. Accessed on July 3, 2014.

Easton, Jeff. "Fracking Wastewater Management." WaterWorld, http://www.waterworld.com/articles/wwi/print/volume-28/issue-5/regional-spotlight-us-caribbean/fracking-wastewater-management.html. Accessed on June 27, 2014.

Ellsworth, William L. 2013. "Injection-Induced Earthquakes," *Science* 341 (6142): 142.

"Energy Dept. Welcomes Fracking Chemical Disclosure." 2014. Fuel Fix, April 25. http://fuelfix.com/blog/2014/04/25/energy-dept-welcomes-fracking-chemical-disclosure/. Accessed on June 28, 2014.

"Environmental Law." Legal Information Institute, http://www.law.cornell.edu/wex/environmental_law. Accessed on July 3, 2014.

"Evaluation of Impacts to Underground Sources of Drinking Water by Hydraulic Fracturing of Coalbed Methane Reservoirs; National Study Final Report." 2004. United States Environmental Protection Agency, June. http://www.epa.gov/ogwdw/uic/pdfs/cbmstudy_attach_uic_final_fact_sheet.pdf. Accessed on July 3, 2014.

Evans, David M. 1966. "The Denver Area Earthquakes and the Rocky Mountain Arsenal Disposal Well," *The Mountain Geologist* 3 (1): 23–36.

Everley, Steve. 2013a. "Bombshell Study Confirms Low Methane Leakage from Shale Gas." Energy in Depth, September 16. http://energyindepth.org/national/bombshell-study-confirm-slow-methane-leakage-from-shale-gas/. Accessed on June 29, 2014.

Everley, Steve. 2013b. "How Anti-Fracking Activities Deny Science: Air Emissions." Energy in Depth, August 12. http://energyindepth.org/national/how-anti-fracking-activists-deny-science-air-emissions/. Accessed on June 26, 2014.

Everley, Steve. 2013c. "How Anti-Fracking Activities Deny Science: Water Contamination." The Energy Collective, August 17.http://theenergycollective.com/saeverley/260556/how-anti-fracking-activists-deny-science-water-contamination. Accessed on June 28, 2014.

"Explore Shale." 2011. Penn State Public Broadcasting, http://exploreshale.org/. Accessed on June 27, 2014.

"Fact Sheet: Economic Impacts of High-Volume Hydraulic Fracturing in New York State." New York State Department of Environmental Conservation, http://www.dec.ny.gov/docs/materials_minerals_pdf/econimpact092011.pdf. Accessed on June 26, 2014.

"Find a Well." FracFocus, http://www.fracfocusdata.org/DisclosureSearch/. Accessed on June 28, 2014.

"Frac Attack: Risks, Hype, and Financial Reality of Hydraulic Fracturing in the Shale Plays." 2010. Reservoir Research Partners and Tudor Pickering Holt and Company, July 8. http://www.oilandgasbmps.org/docs/GEN231-Frac-Attack.pdf. Accessed on June 27, 2014.

"Fracking Caused Hundreds of Complaints about Contaminated Water in 4 States." 2014. RT, January 6. http://rt.com/usa/fracking-chemicals-found-well-water-243/. Accessed on June 28, 2014.

"Fracking Regulations by State." ALS, http://www.alsglobal.com/en/Our-Services/Life-Sciences/Environmental/Capabilities/North-America-Capabilities/USA/Oil-and-Gasoline-Testing/Oil-and-Gas-Production-and-Midstream-Support/Fracking-Regulations-by-State. Accessed on July 3, 2014.

"Fueling the Future with Natural Gas: Bringing It Home." American Gas Association, http://www.aga.org/our-issues/playbook/Documents/2014AGA_Playbook_FINAL.pdf. Accessed on June 25, 2014.

"Gas Goes Boom." 2012. The Economist, http://www.economist.com/node/21556291. Accessed on June 24, 2014.

Gleeson, John. 2011. "Fracking Grievances Aired at Eagle Hill." Mountain View Gazette, September 13. http://www.mountainviewgazette.ca/article/20110913/MVG0801/309139989/0/mvg. Accessed on July 2, 2014.

Golden, Mark. "America's Natural Gas System Is Leaky and in Need of a Fix, New Study Finds." Stanford News, http://news.stanford.edu/news/2014/february/methane-leaky-gas-021314.html. Accessed on June 26, 2014.

Gottesdiener, Ben. "Attempts at Independence." Washington University Political Review, http://www.wupr.org/2014/03/15/attempts-at-independence-2/. Accessed on June 24, 2014.

Green, Mark. "About Monterey Shale . . . Don't Bet Against Innovation." Energy Tomorrow, http://www.energytomorrow.org/blog/2014/may/about-monterey-shale-don't-bet-against-innovation. Accessed on June 23, 2014.

"The Halliburton Loophole." Earthworks, http://www.earthworksaction.org/issues/detail/inadequate_regulation_of_hydraulic_fracturing#.U7XA-vldV8E. Accessed on July 3, 2014.

Hansen, Evan, Dustin Mulvaney, and Meghan Betcher. 2013. "Water Resource Reporting and Water Footprint from Marcellus Shale Development in West Virginia and Pennsylvania." Downstream Strategies, October 30. http://www.downstreamstrategies.com/documents/reports_publication/marcellus_wv_pa.pdf. Accessed on June 27, 2014.

Hansen, Lee R. "Transport, Storage, and Disposal of Fracking Waste." OLR Research Report, http://www.cga.ct.gov/2014/rpt/2014-R-0016.htm. Accessed on June 27, 2014.

Hassett, Kevin A. 2013. "Benefits of Hydraulic Fracturing." American Enterprise Institute, April 4. http://www.aei.org/article/economics/benefits-of-hydraulic-fracking/. Accessed on June 26, 2014.

Heinberg, Richard. "Snake Oil: Chapter 5—The Economics of Fracking. Who Benefits?" Resilience, Accessed on June 25, 2014.

"High-Volume Horizontal Hydraulic Fracturing." 2012. City of Anna, October 3. http://documents.foodandwaterwatch.org/doc/Frack_Actions_AnnaIL.pdf#_ga=1.17517754.229352318.1403631732. Accessed on July 4, 2014.

Holland, Austin. 2011. "Examination of Possibly Induced Seismicity from Hydraulic Fracturing in the Eola Field, Garvin County, Oklahoma." Oklahoma Geological Survey, August. http://www.ogs.ou.edu/pubsscanned/openfile/OF1_2011.pdf. Accessed on July 1, 2014.

"How Much Water Does It Take to Frack a Well?" 2013. Pennsylvania: Energy. Environment. Economy, March 12. http://stateimpact.npr.org/pennsylvania/2013/03/12/how-much-water-it-takes-to-frack-a-well/. Accessed on June 27, 2014.

Howarth, Robert W., Renee Santoro, and Anthony Ingraffea. 2011. "Methane and the Greenhouse-Gas Footprint of Natural Gas from Shale Formations. A Letter." *Climatic Change* 106 (4): 679690.

Hughes, J. David. 2013. "Drilling California: A Reality Check on the Monterey Shale." Santa Rosa, CA: Post Carbon Institute. Available online at http://www.postcarbon.org/reports/Drilling-California_FINAL.pdf. Accessed on June 23, 2014.

Hunt, Spencer. 2012. "Is There Enough Water for 'Fracking' Boom?" The Columbus Dispatch, November 27. http://www.dispatch.com/content/stories/local/2012/11/27/is-there-enough-water-for-fracking-boom.html. Accessed on June 27, 2014.

"Hydraulic Fracturing Chemical Disclosure Policy." Baker Hughes, http://public.bakerhughes.com/shalegas/disclosure.html. Accessed on June 28, 2014.

"Hydraulic Fracturing 101." Earthworks, http://www.earthworksaction.org/issues/detail/hydraulic_fracturing_101#.U67z-fldV8E. Accessed on June 28, 2014.

"Induced Earthquakes." 2014. U.S. Geological Survey, January 16. http://earthquake.usgs.gov/research/induced/. Accessed on July 1, 2014.

J-127A-D-2012. [Robinson Township, et al. *vs. Commonwealth of Pennsylvania*] December 19, 2013. http://www

.pacourts.us/assets/opinions/Supreme/out/J-127-AD-2012 oajc.pdf?cb=1. Accessed on July 5, 2014.

"Keep Tap Water Safe." http://keeptapwatersafe.org/global -ban-son-fracking/. Accessed on July 4, 2014.

Kelly, Sharon. "Shale Rush Hits Argentina as Oil Majors Spend Billions on Fracking in Andes Region." Desmogblog.com, http://www.desmogblog.com/2014/05/29/shale-drilling -rush-hits-argentina-oil-majors-spend-billions-fracking -andes. Accessed on June 24, 2014.

Kiger, Paul J. 2014a. 2014. "Scientists Warn of Quake Risk from Fracking Operations." National Geographic Daily News, May 2. http://news.nationalgeographic.com/news/ energy/2014/05/140502-scientists-warn-of-quake-risk-from -fracking-operations/. Accessed on July 2, 2014.

Kiger, Paul J. 2014b. 2014. "Simple Fixes Could Plug Methane Leaks from Energy Industry, Study Finds." National Geographic, March 19. http://energyblog.national geographic.com/2014/03/19/simple-fixes-could-plug-methane -leaks-from-energy-industry-study-finds/. Accessed on June 29, 2014.

Kim, Won-Young. 2013. "Induced Seismicity Associated with Fluid Injection into a Deep Well in Youngstown, Ohio," *Journal of Geophysical Research: Solid Earth* 118 (7): 3506–3518.

Klare, Michael. 2012. "U.S. Energy Independence Is a Pipe Dream." Salon, October 4. http://www.salon.com/2012/10/ 04/u_s_energy_independence_is_a_pipe_dream/. Accessed on June 25, 2014.

Koontz, Heidi. 2013. "Earthquake Swarm Continues in Central Oklahoma." U.S. Geological Survey, October 22. http://www.usgs.gov/newsroom/article.asp?ID=3710& from=rss#.U7NAxPldV8E. Accessed on July 1, 2014.

Legal Environmental Assistance Foundation, Inc., Petitioner, vs. United States Environmental Protection Agency, Respondent.

118 F.3d 1467 (1997). https://law.resource.org/pub/us/case/reporter/F3/118/118.F3d.1467.95-6501.html. Accessed on July 3, 2014.

Lester, Kerry. 2013. "Ill. Passes Nation's Toughest Fracking Regulations." AP, June 1. http://bigstory.ap.org/article/ill-passes-nations-toughest-fracking-regulations. Accessed on July 3, 2014.

Levi, Michael. 2013a. "The Emerging Irony of US Energy Independence." Megatrends, http://www.cnbc.com/id/100780504?__source=yahoo%7Cfinance%7Cheadline%7Cheadline%7Cstory&par=yahoo&doc=100780504%7CThe%20Emerging%20Irony%20of%20US. Accessed on June 25, 2014.

Levi, Michael. 2013b. *The Power Surge: Energy, Opportunity, and the Battle for America's Future*. Oxford, UK: Oxford University Press.

"List of Bans Worldwide." Keep Tap Water Safe, http://keeptapwatersafe.org/global-ban-son-fracking/. Accessed on June 24, 2014.

Loomis, David G. 2012. "The Potential Economic Impact of New Albany Gas on the Illinois Economy." December. http://news.heartland.org/sites/default/files/illinois_fracking_study.pdf. Accessed on June 26, 2014.

"Man-Made Earthquakes Update." 2014. U.S. Geological Survey, January 17. http://www.usgs.gov/blogs/features/usgs_top_story/man-made-earthquakes/. Accessed on July 2, 2014.

Menenberg, Aaron. 2012. "Let's Get Real: Energy Independence Is an Unrealistic and Misleading Myth." EconoMonitor, September. http://www.economonitor.com/policiesofscale/2012/09/06/lets-get-real-energy-independence-is-a-unrealistic-and-misleading-myth/. Accessed on June 24, 2014.

"Methane Emissions from the Natural Gas Industry." 1996. U.S. Environmental Protection Agency, June. http://www.epa.gov/gasstar/documents/emissions_report/2_technicalreport.pdf. Accessed on June 29, 2014.

Miller, Scott M., et al. 2013. "Anthropogenic Emissions of Methane in the United States." *Proceedings of the National Academy of Sciences of the United States of America* 110 (50): 20018–22.

"Monterey Shale Downgraded." Post Carbon Institute, http://www.postcarbon.org/press-release/2239062-government-slashes-california-oil-forecast. Accessed on June 23, 2014.

Morris, Jim, Lisa Song and David Hasemyer. 2014. "Big Oil, Bad Air." The Center for Public Integrity, February 18. http://eagleford.publicintegrity.org/. Accessed on June 30, 2014.

Mufson, Steven. 2014. "How Two Small New York Towns Have Shaken up the National Fight over Fracking." The Washington Post, July 2. http://www.washingtonpost.com/business/economy/how-two-small-new-york-towns-have-shaken-up-the-national-fight-over-fracking/2014/07/02/fe9c728a-012b-11e4-8fd0-3a663dfa68ac_story.html. Accessed on July 5, 2014.

Muller, Richard A., and Elizabeth A. Muller. 2013. "Why Every Serious Environmentalist Should Favour Fracking." Centre for Policy Studies, December. http://www.cps.org.uk/files/reports/original/131202135150-WhyEverySeriousEnvironmentalistShouldFavourFracking.pdf. Accessed on June 26, 2014.

Murphy, Dylan, and Jo Murphy. 2014. "Fracking the U.S.—Monterey Shale's 96% Downgrade Blows the Scam." Ecologist, http://www.theecologist.org/News/news_analysis/2442482/fracking_the_us_monterey_shales_96_downgrade_blows_the_scam.html. Accessed on June 22, 2014.

Mushalik, Matt. 2004. "World Crude Production 2013 without Shale Oil Is Back to 2005 Levels." Resilience, http://www.resilience.org/stories/2014-03-26/world-crude-production-2013-without-shale-oil-is-back-to-2005-levels. Accessed on June 24, 2014.

"9 Countries or Regions That Ban Fracking." Petro Global News, http://petroglobalnews.com/2013/10/9-countries-or-regions-that-ban-fracking/. Accessed on June 24, 2014.

"North America Leads the World in Production of Shale Gas." U.S. Energy Information Administration, http://www.eia.gov/todayinenergy/detail.cfm?id=13491. Accessed on June 24, 2014.

"Nuisance." The Free Dictionary, http://legal-dictionary.thefreedictionary.com/nuisance. Accessed on July 2, 2014.

Ordinance #2013-01. State of New Mexico. County of Mora, http://documents.foodandwaterwatch.org/doc/Frack_Actions_MoraCountyNM.pdf#_ga=1.106642432.229352318.1403631732. Accessed on July 4, 2014.

Ordinance #2014-006. City of Canandaigua, http://documents.foodandwaterwatch.org/doc/Frack_Actions_CanandaiguaNY.pdf#_ga=1.38032160.229352318.1403631732. Accessed on July 4, 2014.

Ordinance #659. 2011. Borough of West Homestead, May 10. http://documents.foodandwaterwatch.org/doc/Frack_Actions_WestHomesteadPA.pdf#_ga=1.218903450.229352318.1403631732. Accessed on July 4, 2014.

"Overview of Greenhouse Gases." U.S. Environmental Protection Agency. http://epa.gov/climatechange/ghgemissions/gases/ch4.html. Accessed on June 26, 2014.

"PA Legislature Pre-empts Communities on Fracking." 2012. Community Environmental Legal Defense Fund, March. http://www.celdf.org/section.php?id=343. Accessed on July 5, 2014.

Pacala, S., and R. Socolow. 2004. "Stabilization Wedges: Solving the Climate Problem for the next 50 Years with Current Technologies." *Science* 305 (5686): 968–972.

"Particulate Matter." U.S. Environmental Protection Agency, http://www.epa.gov/airquality/particulatematter/index.html. Accessed on June 30, 2014.

"Pennsylvania Alliance for Clean Water and Air." June 7, 2014. http://pennsylvaniaallianceforcleanwaterandair.wordpress.com/the-list/. Accessed on June 28, 2014.

Pétron, Gabrielle, et al. 2012. "Hydrocarbon Emissions Characterization in the Colorado Front Range: A Pilot Study." *Journal of Geophysical Research* 117 (D4), http://onlinelibrary.wiley.com/doi/10.1029/2011JD016360/abstract. Accessed on June 29, 2014.

Porter, David J. 2014. "U.S. House of Representatives Committee on Science, Space, and Technology: Examining the Science of EPA Overreach: A Case Study in Texas," February 5. http://www.rrc.state.tx.us/media/1012/epaoverreach.pdf. Accessed on June 28, 2014.

Raichel, Daniel. 2013. "HUGE Victory!: PA Supreme Court Says State Can't Force Fracking on Local Communities." Natural Resources Defense Council Staff Blog, December 20. http://switchboard.nrdc.org/blogs/draichel/huge_victory_in_pennsylvania_p.html. Accessed on July 5, 2014.

Redden, Jim. 2012. "Unlocking the Secrets of the U.S.' Largest Onshore Oil Reserves." *World Oil*, November: 88–96. Also available online at http://www.slb.com/~/media/Files/industry_challenges/unconventional_gas/industry_articles/20133001_world_oil_monterey.pdf. Accessed on June 22, 2014.

"Report: Methane Emissions at 'High Levels' in Uintah Basin." 2013. NGI's Shale Daily. August 6. http://www.naturalgasintel.com/articles/643-report-methane-emissions

-at-high-levels-in-uintah-basin. Accessed on June 29, 2014.

"Review of Emerging Resources: U.S. Shale Gas and Shale Oil Plays." U.S. Energy Information Administration, http://www.eia.gov/analysis/studies/usshalegas/pdf/usshaleplays.pdf. Accessed on June 22, 2014.

Ridlington, Elizabeth, and John Rumpler. 2013. "Fracking by the Numbers: Key Impacts of Dirty Drilling at the State and National Level." October. Boston: Environment America. Available online at http://www.environmentamerica.org/sites/environment/files/reports/EA_FrackingNumbers_scrn.pdf. Accessed on June 24, 2014.

Rogers, Deborah. 2013. "Shale and Wall Street: Was the Decline in Natural Gas Prices Orchestrated?" Energy Policy Forum, February. Available online at http://shalebubble.org/wp-content/uploads/2013/02/SWS-report-FINAL.pdf. Accessed on June 25, 2014.

Rühl, Christof. 2013. "Oil Boom 2.0—An American Dream Updated." https://www.linkedin.com/today/post/article/20130730080645-259060403-oil-boom-2-0-an-american-dream-updated. Accessed on June 23, 2014.

Rushton, Lisa, and Candice Castaneda. 2014. "Drilling Into Hydraulic Fracturing and the Associated Wastewater Management Issues." Paul Hastings. Stay Current, May. http://www.paulhastings.com/docs/default-source/PDFs/stay-current-hydraulic-fracturing-wastewater-management.pdf. Accessed on July 2, 2014.

Sahagun, Louis. 2014. "U.S. Officials Cut Estimate of Recoverable Monterey Shale Oil by 96%." Los Angeles Times, http://www.latimes.com/business/la-fi-oil-20140521-story.html. Accessed on June 22, 2014.

"Shale Gas: A Global Perspective." 2011. KPMG Global Energy Institute, December. http://www.kpmg.com/Global/en/IssuesAndInsights/ArticlesPublications/

Documents/shale-gas-global-perspective.pdf. Accessed on June 24, 2014.

"Shale Oil: The Next Revolution." 2013. Shale Oil, February. http://www.pwc.com/en_GR/gr/surveys/assets/shale-oil.pdf. Accessed on June 24, 2014.

"Shale Oil and Shale Gas Resources Are Globally Abundant." U.S. Energy Information Administration, http://www.eia.gov/todayinenergy/detail.cfm?id=14431. Accessed on June 24, 2014.

Smith, Grant. 2013. "U.S. to Be Top Oil Producer by 2015 on Shale, EIA Says." Bloomberg News, http://www.bloomberg.com/news/2013-11-12/u-s-nears-energy-independence-by-2035-on-shale-boom-iea-says.html. Accessed on June 22, 2014.

Smith, Taylor. 2012. "Research & Commentary: Hydraulic Fracturing and Air Quality." The Heartland Institute, September 10.http://heartland.org/policy-documents/research-commentary-hydraulic-fracturing-and-air-quality. Accessed on June 30, 2014.

Song, Lisa. 2012. "First Study of Its Kind Detects 44 Hazardous Air Pollutants at Gas Drilling Sites." Inside Climate News, December 3. http://insideclimatenews.org/news/20121203/natural-gas-drilling-air-pollution-fracking-colorado-methane-benzene-endocrine-health-NMHC-epa-toxic-chemicals. Accessed on June 30, 2014.

Taylor, Kate, and Thomas Kaplan. 2014. "New York Towns Can Prohibit Fracking, State's Top Court Rules." New York Times, June 30. http://www.nytimes.com/2014/07/01/nyregion/towns-may-ban-fracking-new-york-state-high-court-rules.html. Accessed on July 5, 2014.

"Technically Recoverable Shale Oil and Shale Gas Resources: An Assessment of 137 Shale Formations in 41 Countries Outside the United States." 2013. U.S. Energy Information Administration, June. http://www.eia.gov/analysis/studies/

worldshalegas/pdf/fullreport.pdf. Accessed on June 24, 2014.

Tollefson, Jeff. 2012. "Air Sampling Reveals High Emissions from Gas Field: Methane Leaks during Production May Offset Climate Benefits of Natural Gas." Nature News & Comment. February 7. http://www.nature.com/news/air-sampling-reveals-high-emissions-from-gas-field-1.9982. Accessed on June 29, 2014.

"TOXNET Databases." U.S. National Library of Medicine. http://toxnet.nlm.nih.gov/. Accessed on June 28, 2014.

"2013 State Unemployment Rates." National Conference of State Legislatures, http://www.ncsl.org/research/laborandemployment/2013-state-unemployment-rates.aspx. Accessed on June 25, 2014.

"Underground Injection Control Program." 2013. U.S. Environmental Protection Agency, November 19. http://water.epa.gov/type/groundwater/uic/. Accessed on July 1, 2014.

[Untitled document]. Town of Penn Yan, New York. http://documents.foodandwaterwatch.org/doc/Frack_Actions_PennYanNY.pdf#_ga=1.10170422.229352318.1403631732. Accessed on July 4, 2014.

"U.S. Crude Oil Proved Reserves." U.S. Energy Information Administration, http://www.eia.gov/dnav/pet/hist/LeafHandler.ashx?n=PET&s=RCRR01NUS_1&f=A. Accessed on June 22, 2014.

"U.S. Natural Gas, Wet after Lease Separation Proved Reserves." U.S. Energy Information Administration, http://www.eia.gov/dnav/ng/hist/rngr21nus_1a.htm. Accessed on June 22, 2014.

"US 'Energy Independence': Geopolitical Consequences." 2013. Business Monitor News & Views, February. http://www.businessmonitor.com/news-and-views/us-energy-independence-geopolitical-consequences. Accessed on June 24, 2014.

"Village Council Resolution No. 2012-614." 2012. Meyers Lake Village, April 9. http://documents.foodandwaterwatch.org/doc/Frack_Actions_MeyersLakeOH.pdf#_ga=1.117659018.229352318.1403631732. Accessed on July 4, 2014.

"Wastewater Disposal May Trigger Quakes at a Greater Distance than Previously Thought." 2014. Seismological Society of America, May 1. http://www.seismosoc.org/society/press_releases/SSA_2014_Induced_Seismicity_Press_Release.pdf. Accessed on July 2, 2014.

Weinhold, Bob. 2012. "The Future of Fracking." *Environmental Health Perspectives* 120 (7): A272–A279. Also available online at http://ehp.niehs.nih.gov/wp-content/uploads/120/7/ehp.120-a272.pdf. Accessed on June 30, 2014.

Woodrum, Amanda. 2014. "Fracking in Carroll County, Ohio." Policy Matters Ohio, April. http://www.policymattersohio.org/wp-content/uploads/2014/04/Shale_Apr2014.pdf. Accessed on July 2, 2014.

Woods, Josh. 2012. "Truck Dust, Noise and Fumes Still a Concern in Sandy." http://www.theprogressnews.com/default.asp?read=30940. Accessed on July 2, 2014.

Worstall, Tim. 2014. "If We Judged the Monterey Shale by Green Energy Standards Then the 96% Cut in Recoverable Oil Is Great News." Forbes, May 22. http://www.forbes.com/sites/timworstall/2014/05/22/if-we-judged-the-monterey-shale-by-green-energy-standards-then-the-96-cut-in-rcoverable-oil-is-great-news/. Accessed on June 23, 2014.

Yergin, Daniel. 2013. "Congratulations, America. You're (Almost) Energy Independent. Now What?" Politico Magazine, http://www.politico.com/magazine/story/2013/11/congratulations-america-youre-almost-energy-independent-now-what-98985.html#.U6n2LPldV8E. Accessed on June 24, 2014.

Introduction

Fracking is a complex and controversial issue. Many technical, political, social, economic, environmental, and other questions are involved in decisions as to where, when, and how drilling for oil and gas is to be conducted. This chapter provides interested observers with an opportunity to express their views on this range of topics related to hydraulic fracturing.

Separating Fact from Hype: Trudy E. Bell

Whenever fortunes and political reputations stand to be made by a new technology, but votes and political support are needed from John and Jane Q. Citizen for permits and permissions before it can proceed, take heed: Before any town meetings, do some serious homework to bone up on actual facts and to ask essential questions. Informed independent thinking is

---

Workers move a section of well casing into place at a Chesapeake Energy natural gas well site near Burlington, Pennsylvania, in Bradford County. So vast is the wealth of natural gas locked into dense rock deep beneath Pennsylvania, New York, West Virginia and Ohio that some geologists estimate it's enough to supply the entire East Coast for 50 years. (AP Photo/Ralph Wilson)

especially crucial if the technology is surrounded by a swirl of controversial claims and counterclaims. How can a regular person begin to figure out what's the truth and whose claims to trust?

Examine all sides of all claims, asking whether the claims are complete and accurate. Seek to determine who is making the claims, the constituencies those individuals or organizations represent, and who is financing them. Such a deep dive into who-says-what-and-why is essential when controversies become heated, because opposing groups often hire public relations experts to present their viewpoints in the most favorable light. Some PR folk may even be "spin doctors," slang for experts in spinning a network of half-truths, often using tactics that appeal to emotions rather than to intellect and reason. Thus, you also need to satisfy yourself whether adversaries are fighting fair—that is, bolstering their cases based on objective, verifiable facts and peer-reviewed scientific data, or whether they are fighting dirty—that is, relying on swaying public opinion through name-calling or casting doubt on the competence or assertions of the opposition.

Four fundamental sets of cost-benefit questions can get you started. Note: Even though the examples below are tailored to horizontal drilling for petroleum products, such cost-benefit questions are relevant for exploring the pros and cons of any large-scale engineering project, so file them away for future reference. In general, it is wise to remain skeptical and do diligent research rather than rush forward uninformed. Large-scale engineering projects have many unexpected complexities and take years to complete—and once started, rarely is it possible to turn back the clock and return to pre-project status.

1. What are the claimed benefits? For whom? How are the benefits measured? Who else might benefit, even if they are not publicly claimed to be beneficiaries?

   Many claims are made in favor of fracking—and claims may vary around the country. In the Utica and Marcellus shale regions of Ohio and Pennsylvania, where horizontal drilling is primarily for natural gas (methane), for example, petroleum companies emphasize that methane is a lower-carbon fuel than coal and oil and its combustion emits less of the greenhouse gas carbon dioxide (CO_2), and thus converting from coal or oil to natural gas can help reduce harmful CO_2 emissions that drive climate change.

That is true—but only part of the story. In the Bakken shale region of North Dakota, the horizontal drilling is primarily for oil. Indeed, the methane that naturally bubbles up with the oil is either vented or flared (burned off) at the well site, because natural gas sells for so much less on the world market than oil that petroleum companies choose not to invest in building infrastructure and pipelines for capturing the methane for sale. The amount of natural gas thus wasted is significant: nearly a third of North Dakota's annual production, and enough to heat every U.S. household for three days. Moreover, the methane vented into the air is a greenhouse gas some 25 times more potent than CO_2.

Advertising by the petroleum companies often emphasizes how horizontal drilling for oil is helping to free the country from dependence on foreign oil—even making the United States a net exporter of oil. Also emphasized is the possibility of the technology creating new jobs at home for the drilling as well as for a host of supporting industries (trucking, road building, housing for workers, restaurants, etc.). Less publicized—excepting in programs covering business news—is the profitability to the petroleum companies, making their owners and stockholders the biggest beneficiaries of horizontal drilling.

The ultimate temptation held out to local homeowners is the potential of personally growing rich if oil or gas is found under your property and you sell or lease your mineral rights to drillers in exchange for being paid royalties (a percentage of the profits). To be sure, some individual homeowners have become rich through leasing or selling their mineral rights. Others, however, have discovered that their ownership of their land does not include the mineral rights, yet they are compelled to allow drillers access to their property (a legal situation known as a *split estate*). Others may not want to sell or lease their mineral rights, but are nonetheless compelled to do so if most of their neighbors do (a legal situation known as *forced pooling*).

2. What are the claimed risks or dangers? Do the beneficiaries share the risks? Do the risk takers share the benefits? What are the sources of information?

   Residents in a number of states have complained that methane gas or oily substances or other chemicals have appeared in drinking water pumped from local wells near the sites of horizontal drilling, causing headaches or other symptoms, and have filed lawsuits to stop local drilling and redress such alleged pollution. Petroleum companies—whose owners, stockholders, or employees may live hundreds of miles away from the drilling sites—have strongly contested assertions that the process of hydraulic fracturing a horizontal well can cause chemicals or petroleum products to enter groundwater aquifers, although they have acknowledged that defective seals around the upper part of wells could allow some leakage. It has been difficult for teams of scientists, environmental organizations, or legal experts to determine the objective facts because in many cases baseline measurements of water quality were not made before drilling started; moreover, once drilling started, some companies have prohibited sufficient access to drilling sites for taking ongoing measurements. Some states concerned about their water supplies are considering or have passed new regulations mandating that chemical composition of drinking water aquifers be tested regularly before, during, and after drilling.

   Significant earthquakes in the Midwest and Great Plains, regions not seismically active, have been widely documented from shale gas activities by geologists writing in peer-reviewed journals. The culprit is not the horizontal drilling or hydraulic fracturing itself, but the high-pressure disposal of the wastewater that flows back out of the well after it is hydraulically fractured. Because the wastewater is far saltier than seawater and also contains naturally occurring radioactive materials, it cannot be purified for reuse by regular municipal water treatment plants, nor discharged into local lakes or rivers. So it is discarded by being injected under high pressure miles deep underground into Class II injection wells. This high-pressure wastewater, however, can "lubricate" dormant earthquake faults deep underground, causing slippage and triggering earthquakes: one in Oklahoma was measured to be a damaging 5.7 on the moment magnitude scale (which has replaced the Richter scale). That has raised particular concern among residents of seismically active California, where petroleum companies seek to drill the Monterey shale.

3. Is any part of the controversy framed in terms of false dichotomies? Are there other methods of achieving the claimed benefits? If so, what are the pros and cons of the proposed technology compared to other competing methods?

   Sometimes a controversy gets framed in such a way that it appears that technology proponents are pro jobs whereas technology opponents care only about trees or animals instead of people. In such cases, dig deeper to ascertain the actual concerns. People living near drilling sites need clean water to drink and clean air to breathe, and so do the cattle and other livestock they may be raising. Moreover, farmers' and ranchers' livelihoods depend on their crops and livestock not being exposed to environmental toxins that either cause illness or introduce chemicals into the food chain. Last, in certain parts of the country, wind energy or solar energy may be economically competitive as well as cleaner (zero CO_2 emissions) and the source of new high-tech jobs.

4. Where benefits or risks are quantified primarily in terms of money, is money everything? What else might be

important, especially for the long term (your children and grandchildren)?

Many states are blessed with natural wonders or natural resources—including trophy hunting in forests and wetlands, trout fishing in pristine rivers, breathtaking scenery, and natural peace and quiet—that draw tourism from around the world, accounting for a significant part of the state economy. Yet, permits for horizontal drilling—with its concomitant roads, truck traffic, 24/7 stadium lighting while drilling, and chemicals—have been sought under national parklands (including the Grand Canyon). That has caused some far-sighted state legislators and other stewards of natural resources to reflect on the potential impact of the technology for decades to come, long after the drillers have departed, and ask: What unanticipated surprises might we inherit? Far beyond the next quarter's profits, what choices are wisest for the long term?

Trudy E. Bell, M.A., a former editor for Scientific American *and* IEEE Spectrum *magazines, has written, coauthored, or edited a dozen books and several hundred articles. She is a senior writer for the University of California High-Performance Astro-Computing Center (UC-HiPACC) and has edited seven reports for the Union of Concerned Scientists.*

Why Fracking Is Beneficial: Bruce Everett

The term *fracking* generally refers to the production of natural gas by the hydraulic fracturing of source rocks deep underground. This technology has revolutionized the American energy market in the last few years. Before we discuss the specifics of this technology and its pros and cons, let's start with some context.

The development of humanity, and in particular the creation of modern industrial society, is based largely on the substitution of chemical and mechanical energy for human and animal labor. The discovery and widespread use of fossil fuels (coal, oil, and natural gas) in the eighteenth and nineteenth centuries transformed the United States and Europe from a collection of largely poor, rural communities into mobile, industrial nations whose citizens enjoy unprecedented wealth, mobility, freedom, and knowledge.

Although energy brings us enormous benefits, every form of energy has its drawbacks. Coal, for example, contains many impurities, like sulfur and heavy metals, which pollute the air when it is burned. Nuclear energy produces virtually no pollution in normal operation but presents serious challenges in producing, handling, and disposing of the highly radioactive materials which power atomic reactors. Oil, our only viable transportation fuel, is concentrated in unstable countries, creating difficult national security problems.

In recent decades, many people have placed their hope in renewable energy, mainly solar and wind, to meet our energy needs without suffering the drawbacks of fossil fuels. Although wind and solar energy are clean and abundant, these technologies remain way too expensive for widespread use and perform poorly. In particular, unlike fossil fuels, nature gives us wind and solar energy when it wants to, not when we need it.

Through all our energy debates of the last 40 years, natural gas has sat quietly in the background. We have long recognized the advantages of natural gas, in particular its cleanliness and convenience for use in homes and power plants, but the United States was thought to have only modest natural gas resources, and its high price limited its ability to compete with coal in power generation.

In the last few years, natural gas has taken a more prominent place in our energy supply because of hydraulic fracturing. In conventional natural gas reservoirs, a source rock, like sandstone, contains natural gas trapped in the rock thousands of feet underground. When a drill penetrates the rock, the natural gas flows through the permeable structure of the rock and up the well to the surface. The United States, however, also has enormous amounts of natural gas trapped in tight rock structures which do not permit the gas to flow easily. Fracking these tight reservoirs involves drilling horizontally through the rock structure, rather than just vertically. High-pressure water is then pumped into the well, which cracks the rock allowing channels for the gas to flow into the open well. Sand and small amounts of chemicals are also pumped into the well to keep the cracks open.

This new technology has opened up large new natural gas resources which are inexpensive to produce, and located not only in traditional oil and gas states, like Texas, but in areas like Pennsylvania, which are much closer to the big natural gas markets of the Northeast.

Because of its availability and low cost, fracking has brought a dramatic reduction in the price of natural gas. In May 2008, the price of natural gas at the Henry Hub market in Louisiana reached nearly $11.50 per thousand cubic feet (MCF) of natural gas (Natural Gas Spot and Futures Prices 2014). By July of 2014, the price had fallen to about $4.25/MCF—a drop of nearly two-thirds (Natural Gas and Spot and Futures Prices 2014). Imagine, for example, the price of gasoline falling from its current level of $3.70 per gallon to $1.35.

The enhanced availability of low-priced natural gas has brought many economic and environmental benefits. Despite continued growth in power demand, electricity prices in the United States remain stable and are only about half the price paid by household consumers in Europe or Japan (Electricity Prices for Households of Selected Countries 2014).

Lower energy costs also enhance U.S. competitiveness and are likely to support a growth in job opportunities in the United States. A recent report by PriceWaterhouseCoopers concluded that fracking could increase U.S. manufacturing employment by one million jobs and save companies $11.6 billion in costs by 2025 (Shale Gas: Reshaping the US Chemicals Industry 2012).

Moreover, low-cost natural gas has allowed us to build fewer coal-fired power plants and even to close some existing coal plants without increasing the cost to consumers. As a result, our air is cleaner without any increase in cost.

As with every other source of energy, fracking raises issues. The process requires a great deal of water, which is circulated through the well and back up to the surface. This water must be dealt with by water treatment, by storage and evaporation, or by reinjection underground. The use of chemicals in the water has raised fears of contamination of local water supplies. In reality, however, sensible government regulation is all that is required to ensure safe water management in fracking operations. Techniques for handling the water are well known, easily applied, and relatively inexpensive.

Some opponents have noted with alarm an increase in earthquake activity around water reinjection operations in states such as Oklahoma. These earthquakes are so small, however, that most cannot even be detected except with the most sensitive instruments, and none has been shown to cause any damage, injury, or disruption.

People in the energy-producing states of the Gulf Coast and Midwest, particularly Texas, Louisiana, and Oklahoma, are used to the impacts of drilling and related activities. For people in Pennsylvania and New York, however, these new industries bring some disruptions, including heavy trucking, noise, and an influx of new people in previously quiet rural towns. Pennsylvania has settled most of these issues through regulation, recognizing that the jobs, royalty payments, and tax revenue from natural gas production far outweigh the problems. In New York, however, the state retains a ban on fracking activities as people argue about the pluses and minuses.

The question to be asked is not whether fracking has its drawbacks—it does. The question is what's the best way to meet our large and growing energy needs, particularly for ever more electricity. Natural gas has major economic and environmental advantages while its downside is quite small. We need to decide how the cost-benefit balance of hydraulic fracturing of natural gas compares with that of other forms of power production, including coal, nuclear, wind, and solar. On balance, fracking looks pretty good.

References

"Electricity Prices for Households of Selected Countries." U.S. Energy Information Administration. http://www.eia.gov/countries/prices/electricity_households.cfm. Accessed on July 14, 2004.

"Natural Gas Spot and Futures Prices." U.S. Energy Information Administration. http://www.eia.gov/dnav/ng/ng_pri_fut_s1_d.htm. Accessed on July 14, 2014.

"Shale Gas: Reshaping the US Chemicals Industry." PWC. October 2012. http://www.pwc.com/en_US/us/industrial-products/publications/assets/pwc-shale-gas-chemicals-industry-potential.pdf. Accessed on July 14, 2014.

Bruce Everett is an adjunct associate professor of international business at the Fletcher School at Tufts University. Before joining the Fletcher faculty in 2003, he held a series of management positions at ExxonMobil Corporation from 1980 to 2002 and served in the United States Department of Energy from 1974 to 1980. He grew up in the Boston area and holds an AB degree from Princeton University and a PhD from the Fletcher School.

The Urgent Need for Global Definition of Terms in the Reporting on Fracking and Seismic Activity: Gina Hagler

Hydraulic fracking is one of the more recent instances of technology potentially run amok. Detractors attribute fouled groundwater and increased seismic activity, to name a few, to this practice of releasing the natural gas inside shale by drilling and then forcing liquid into the ground at high pressure. Advocates of the method contend that fracking has been in practice for more than 60 years, during which the associated environmental impact has been negligible.

For those who are intimately involved in fracking, the debate is shaded with nuance: Which sites are being used as examples? How old are the sites? What safeguards were in use at the start? Have the sites been active since inception? Each question engenders others—ever more meaningful to the participants, while ever more confusing to the casual observer. Yet even the most remote of the casual observers ultimately has a stake in the outcome.

One could argue that fouled water can be treated before consumption and low natural gas prices benefit us all. It is difficult to argue that we can ignore significant seismic activity due to fracking, if that is indeed the case. Even if we don't have personal experience with this byproduct of fracking, we can't sit passively on the sidelines as the earthquakes-due-to-fracking debate heats up. Legislation related to fracking requires a vote by our representatives (Gesing 2014). These representatives want their votes to reflect the wishes of their constituency. Before we can tell them our wishes, we have to know what they are. Before we can decide upon them, we must understand the issues and determine the validity of the data used to support both sides of the argument since—of course—detractors and supporters do not agree on the role of fracking in seismic activity.

To help us with this task of deciding, we could turn to those involved in the debate. We would quickly discover that the experts of one group adamantly refute the assertions of the other. We may sense that the two groups are not arguing over the same precise point, but that nuance is often nearly impossible to trace to an actual fact. Even if we can trace it back, we stand with a trophy of dubious value accompanied by the question of where this new knowledge leaves us. As a result, most of us turn to news coverage, seeking the answer to a simple question: Do we need to be concerned about the role hydraulic fracking may cause in earthquakes?

A recap of recent news stories from sources ranging from science news sites and journals to general news sites reveals that seismic activity and fracking are currently benefitting from active coverage, all of which identifies hydraulic fracking as the cause of seismic activity to one extent or another. From Mother Jones (Sheppard, Lee, and Brownell 2014) to NBC (Are Oklahoma Earthquakes Tied to Fracking? 2014) to NPR (Phillips 2014), seismic activity is attributed to fracking. National Geographic reports that part of the problem is that there is no way to accurately predict which locations will experience tremors (Kiger 2014). All agree the cause of any tremors is not the fracking itself, but the pressurized return of the wastewater to the deep water, underground wells. A review of industry sites reveals their assertions that this return can be done safely.

The back and forth. The slightly different meaning of terms used. The feeling that not quite everyone is showing his or her full hand (U.S. Geological Survey 2014). It is all vaguely reminiscent of environmental issue debates and reporting on such issues as global climate change or the effect of drilling in protected waters. One could argue that this time the outcome of the debate is particularly urgent. This is not a debate about something that can be cleaned up later. There is either seismic activity that leads to an earthquake that causes significant damage, perhaps with loss of life, or there is not. The damage done cannot be undone simply, if at all.

The same vaguely familiar unease can be attributed to a concern that the concerns voiced are overstated: that increased and more sophisticated monitoring has led to the recognition of more events, not because there are truly more, but because they're now on our radar, so to speak (The Facts About Hydraulic Fracturing and Seismic Activity 2014). Even the recent news report that seismic activity worldwide is up leaves one with a lingering doubt as to the role fracking plays—new doubts to accompany the old concern that some of this reporting is more anecdotal than causal (Earthquake Facts and Statistics 2014).

Few things grab public attention as fully as the possibility that the ground beneath our feet will shift and destroy our homes. The role of fracking in seismic activity must be addressed without hype and hysteria (Aho 2014). The media must find a way to assess the true dangers associated with this practice. We must have accurate and responsible coverage by the media—yet how can they deliver that when the involved parties contest the very terms and sites used to define the debate? Without agreement upon the type of site, the age of the site, and the definitions of the process and problem, how can meaningful and transparent reporting occur?

A shared vocabulary and datasets must be put in place. Perhaps major news organizations could advocate for this in their coverage. Perhaps it is the job of our representatives. Most likely it is up to us as individuals to insist on this, since it would ultimately allow a full and unambiguous investigation and understanding of the role of fracking in seismic activity.

References

Aho, Karen. "Earthquakes Are Rising in Oklahoma, and Insurance Is Booming." Bloomberg Businessweek. July 23, 2014. http://www.businessweek.com/articles/2014-07-23/oklahomas-increasing-earthquakes-are-a-boon-for-insurers. Accessed July 25, 2014.

"Are Oklahoma Earthquakes Tied to Fracking?" NBC Nightly News. June 27, 2014. http://www.nbcnews.com/video/nightly-news/55527767#55527767. Accessed June 30, 2014.

"Earthquake Facts and Statistics." U.S. Geological Survey. March 10, 2014. http://earthquake.usgs.gov/earthquakes/eqarchives/year/eqstats.php. Accessed July 25, 2014.

"The Facts about Hydraulic Fracturing and Seismic Activity." American Petroleum Institute. January 1, 2014. http://www.api.org/policy-and-issues/policy-items/hf/facts-about-hydraulic-fracturing-and-seismic-activity.aspx. Accessed on August 6, 2014.

Gesing, Lars. "The Midterm Politics of Fracking." NBC News. June 23, 2014. http://www.nbcnews.com/politics/first-read/midterm-politics-fracking-n138681. Accessed June 24, 2014.

Kiger, Patrick J. "Scientists Warn of Quake Risk from Fracking Operations: Tremors Induced by Wastewater Disposal Are Larger and Harder to Predict than Previously Thought." National Geographic. May 2, 2014. http://news.nationalgeographic.com/news/energy/2014/05/140502-scientists-warn-of-quake-risk-from-fracking-operations/. Accessed June 10, 2014.

Phillips, Susan. "Congressional Watch-Dog Warns Fracking Waste Could Threaten Drinking Water." State Impact: NPR. July 28, 2014. http://stateimpact.npr.org/pennsylvania/2014/07/28/congressional-watch-dog-warns-fracking-waste-could-threaten-drinking-water/. Accessed July 30, 2014.

Sheppard, Kate, Jaeah Lee, and Brett Brownell. "Confirmed: Fracking Triggers Quakes and Seismic Chaos." Mother Jones. July 11, 2013. http://www.motherjones.com/blue-marble/2013/07/earthquakes-triggered-more-earthquakes-near-us-fracking-sites. Accessed July 10, 2014.

U.S. Geological Survey. "Science or Soundbite? Shale Gas, Hydraulic Fracturing, and Induced Earthquakes." YouTube. July 1, 2012. http://youtu.be/vb-BNDx2iIQ. Accessed June 28, 2014.

Gina Hagler is a freelance writer and published author. Her book about fluid dynamic principles in action was published by Springer Verlag in 2013. A member of the NASW and ASJA, she blogs about science, technology, and health at http://www.ginahagler .com.

Fracking and the Future of Fresh Water: Michael Pastorkovich

Life on Earth began in the oceans. All life on Earth requires water in order to survive, prosper, and reproduce. Life deprived of water becomes death. It is absolutely crucial to keep this in mind when discussing the role of water on our planet.

About 71 percent of our planet's surface is covered with water. However, 96.5 percent of all the water on Earth is contained in the highly saline oceans, which is fine for the marine creatures that live in or near the seas and which comprise about half the species in the world. (All data on water are from U.S. Geological Survey 2014). But salt water is poison to the other half, including our own human species, and most of the plants and animals (excepting seafood) we consume for our nourishment. These species, including ours, require fresh water in order to survive.

As fresh water is used up, it is replenished by a delicate natural process called the hydrologic cycle. The heat of the sun evaporates mostly ocean water (leaving the salt behind), which travels through Earth's atmosphere as water vapor until it condenses over land masses in the form of rain, snow, or ice. Heavy, soaking rains seep into the soil and replace water in depleted aquifers.

On the basis of these facts alone, it would seem reasonable to conclude that water conservation ought to be one of humankind's top priorities. But two factors have emerged over the past 100 years which promise to morph into a global water catastrophe by this century's end unless immediate steps are taken to avert the worst of an impending disaster. These two factors are the exponential growth of the human population over the twentieth century and the reality of global warming.

Humanity did not reach the one billion mark until around the year 1800. By 1900 the world's population was still under two billion. Today, the human race stands at seven billion and is rapidly heading towards eight billion. And each of these human beings needs lots of fresh water to stay alive and well and so do most of the plants and animals which each consumes.

But even worse for the world's fresh water supply is the reality of global warming. And let there be no mistake about it: Global warming is real and the principal cause of global warming is the massive infusion of the greenhouse gases carbon dioxide and methane into Earth's atmosphere as a result of human beings' burning fossil fuels like coal, petroleum, and natural gas. This is affirmed by 97 percent of all climate scientists, according to the National Aeronautics and Space Administration and the National Academy of Sciences (Global Climate Change 2014). And according to the World Meteorological Association, the first decade of the twenty-first century was the hottest on record (Past Decade Hottest on Record 2014).

In the face of all this, the most sensible course of action would seem to be to do everything within our power to stabilize the human population (without violating human rights), combat global warming, and conserve every drop of fresh water that we possibly can. However, in the United States today, at least as far as the last two items above go, just the opposite is taking place.

The United States is in the midst of a "shale gas boom." Shale gas is natural gas (mostly methane) that has been locked underground in shale deposits for billions of years. This gas is extracted from the earth by a process called hydraulic fracturing or, in popular parlance, *fracking*. Fracking involves drilling a well shaft into a shale deposit, which usually lies a mile or so beneath the surface of the earth, and then injecting a brew called *fracking fluid*, consisting of huge amounts of fresh water mixed with sand and many chemicals, some toxic, at a very high pressure to crack the shale open and release the gas. The gas is then pumped to the surface.

Fracking has been around for over 60 years but only became economically feasible on a large scale with the perfection of a technique called horizontal drilling, which involves the ability to tilt the drill at a 90 degree angle and bore through the shale for up to a mile in any direction from a single well pad. Previously, conventional vertical drilling required constructing multiple well pads for a single field, which was expensive.

Sometimes shale gas deposits lie beneath aquifers, in which case pipelines are passed through these underground fresh water lakes, presenting the very real possibility of contamination from chemical leakage during the high-pressure fracking procedure. But more outrageous is the fact that each fracking procedure requires more than a million gallons of fresh water (Marcellus Shale 2014). A full 80 percent of that water remains underground, while another 20 percent returns to the surface as a waste product containing not only the dangerous chemicals used in the fracking fluid but also poisonous substances, including heavy metals and radioactive elements, brought up as dissolved rock from Earth's interior itself (Fracking 2014).

Some of this wastewater is stored in on-site open pit *holding ponds*, which can and do develop leaks or overflow in heavy rains, polluting soil and nearby sources of fresh water. But more often this waste is disposed of by the *deep injection* of this poisoned water into subsurface reservoirs such as abandoned gas wells. This method has been shown to cause earth tremors and earthquakes, relatively minor so far (Roach 2014).

Furthermore, wastewater disposed of in this way, like the 80 percent that remains underground in the first place, is removed from the evaporation phase of the hydrologic cycle. The crux is that at least 90 percent of the water used for fracking is ruined forever and rendered unfit for human and animal consumption and for agricultural purposes.

The drilling industry–friendly website EnergyFromShale estimates that at least one million wells have been fracked since the late 1940s (What Is Fracking? 2014). Assuming an average of five million gallons of water per fracking procedure means that 90 percent of five trillion gallons of fresh water have been destroyed and permanently removed from the hyrdrologic cycle. And millions of gallons more are being destroyed every day.

And to what end?

Millions of gallons of fresh water are being destroyed every day in order to extract from the earth fossil fuels, mainly natural gas and crude oil, which, when burned, will add to the plethora of greenhouse gases raising the temperature of our planet and damaging our climate. And, yes, although natural gas burns cleaner than either coal or petroleum, it is a fossil fuel and, hence, does contribute overall to global warming. In fact, according to a study conducted by Stanford University, MIT, and the U.S. Department of Energy, and as reported in *The New York Times*, "the drilling and production of natural gas can lead to leaks of methane, a greenhouse gas 30 times more potent than carbon dioxide" and "those methane leaks negate the climate change benefits of using natural gas as a transportation fuel" (Davenport 2014).

Many parts of the world, including places in the U.S. West are in the midst of a several-years-long drought. Farmers pray for rain and watch their land turn to dust and blow away while watching caravans of trucks go down the highways carrying millions of gallons of water for the frackers to turn into toxic waste. Meanwhile, the Pentagon is preparing for wars and civil conflicts over water around the globe.

There are many reasons to advocate not merely a moratorium but an outright ban on fracking. But the wanton destruction of so much fresh water is the most compelling.

We can live without coal, oil, and natural gas. But we cannot live without water.

References

Davenport, Cora. "Study Finds Methane Leaks Negate Benefits of Natural Gas as a Fuel for Vehicles." The New York Times. February 13, 2014. http://www.nytimes.com/2014/02/14/us/study-finds-methane-leaks-negate-climate-benefits-of-natural-gas.html?_r=0. Accessed on July 18, 2014.

"Fracking." Greenpeace USA. http://www.greenpeace.org/usa/en/campaigns/global-warming-and-energy/The-Problem/fracking/. Accessed on July 18, 2014.

"Global Climate Change." National Aeronautics and Space Administration. http://climate.nasa.gov/. Accessed on July 18, 2014.

"Marcellus Shale." Department of Environmental Conservation. http://www.dec.ny.gov/energy/46288.html. Accessed on July 18, 2014.

"Past Decade Hottest on Record, Marked by Extremes: UN." Fox News. http://www.foxnews.com/world/2013/07/03/past-decade-hottest-on-record-marked-by-extremes-un/. Accessed on July 18, 2014.

Roach, John. "Fracking and Energy Exploration Connected to Earthquakes, Say Studies." NCB News. http://www.nbcnews.com/science/environment/fracking-energy-exploration-connected-earthquakes-say-studies-f6C10604071. Accessed on July 18, 2014.

U.S. Geological Survey. "The USGS Water Science School." http://water.usgs.gov/edu/. Accessed on July 18, 2014.

"What Is Fracking?" EnergyFromShale.com. http://www.energyfromshale.org/hydraulic-fracturing/what-is-fracking. Accessed on July 18, 2014.

Michael Pastorkovich has a degree in philosophy and spent the majority of his working life in the telecommunications industry. He is a long-time member of the Sierra Club and is active in the environmental movement along with other social causes.

Fracking by the Numbers: John Rumpler

In July 2014, a faulty valve atop a railcar owned by FrackMax leaked hydrochloric acid vapors into the air just blocks from downtown Brownsfield, Texas (Hazardous Situation 2014). And two weeks before that, flood-damaged storage tanks spilled 7,500 gallons of crude oil into the Cache la Poudre River, the only river from Colorado listed in the National Wild and Scenic Rivers System (Oil Spill Cleanup Underway in Windsor 2014). And in Ohio, a house with a leaking gas well on the property exploded, killing a 27-year-old woman living there (Investigation Continues at Site of House Explosion 2014).

These are just a few recent examples of the fallout from fracking. Yet alarmingly, this dirty drilling remains exempt from key provisions of our nation's environmental and public health laws.

So as state and federal officials decide whether to allow the oil and gas industry to expand fracking—even to areas that provide drinking water to millions of Americans—it's time to measure the damage done to our environment and health by dirty drilling.

In 2013, we published our findings in a report called "Fracking by the Numbers" (Fracking by the Numbers 2013). We looked at key measures of fracking risks to our water, air, land, and climate.

Contamination of drinking water is one of the key threats posed by fracking. In reviewing state records, we found more than 1,000 documented cases where dirty drilling has contaminated groundwater or other drinking water sources. While such contamination can happen at several points in the fracking process—spills of fracking fluid, well blowouts, leaks around the well bore—perhaps the greatest threat to our water comes from the toxic waste that fracking generates.

Often laced with cancer-causing and even radioactive material, this fracking waste has leaked from waste pits into groundwater, has been dumped into rivers and streams, and has spread onto roadways.

So how much of this fracking waste are we talking about? Using state- and industry-submitted data, we calculated that fracking generated 280 billion gallons of toxic wastewater in 2012 alone. That's enough fracking waste to flood all of Washington, D.C., in a 22-foot-deep toxic lagoon.

And yet, this toxic fracking waste is exempt from our nation's hazardous waste laws.

We also looked at how much water is used by fracking—a crucial question especially in arid Western states in the midst of the fracking frenzy. Again, using data submitted by fracking operators, we calculated that fracking has used 250 billion gallons of fresh water since 2005. And unlike other water uses, to the extent that fracking converts this water into toxic waste that is injected deep down into the ground, water used for fracking is gone for good.

The numbers on fracking look equally appalling with respect to our health, our natural heritage, and the planet.

With a growing number of residents experiencing illness with the onset of nearby fracking operations, we found that fracking operations produced 450,000 tons of air pollution in one year.

To give you a sense of what these air pollution numbers can mean, look at what Pam Judy, a woman whose house in Carmichael, Pennsylvania, is just 780 feet from a compressor station, had to say:

> Shortly after operations began, we started to experience extreme headaches, runny noses, sore/scratchy throats, muscle aches and a constant feeling of fatigue. Both of our children are experiencing nose bleeds and I've had dizziness, vomiting and vertigo to the point that I couldn't stand and was taken to an emergency room. (Personal Account from the Marcellus Shale 2014)

When she finally persuaded state officials to test the air around her house, they found the following chemicals: benzene, styrene, toluene, xylene, hexane, heptane, acetone, acrolein, propene, carbon tetrachloride, and chloromethane. Other residents living on the frontlines of fracking are experiencing similar illnesses, as documented in Shalefield Stories and other publications, as well as reviews by medical doctors.

Fracking is also putting our natural heritage at risk. As pristine landscapes have been covered with well pads, compressor stations, access roads, and other drilling infrastructure, we estimate that fracking has directly degraded 360,000 acres of land since 2005.

And finally, as fracking boosters have sought to win over the green-minded, we calculated that well completions alone—not counting the rest of fracking operations—have produced 100 million metric tons of global warming pollution since 2005. Looking at the full life-cycle of production and use, the global warming impact of dirty gas could well rival that of coal.

Our calculations understate the true toll of fracking. Fracking also inflicts other damage we did not quantify in our report—ranging from contamination of residential wells to ruined roads to earthquakes near disposal sites.

Viewed in their totality, the numbers on fracking add up to an environmental nightmare. Given the number and severity of threats posed by fracking, constructing a regulatory regime sufficient to protect our water and our health—much less enforcing it at more than 80,000 wells, plus processing and waste disposal sites across the country—seems implausible at best. At the end of the day, protecting our environment and public health will require a ban on fracking.

Thus it is no surprise that a growing number of communities across the country—from Dryden, New York, to Dallas, Texas—are taking action to stop dirty drilling within their borders.

And while the power of the oil and gas industry bars such decisive action at the federal level, there are two things President Obama can do right now to at least limit the damage from fracking. First, as the Bureau of Land Management mulls weak rules for fracking on public lands, the president should insist on following a key recommendation of his own administration's advisory panel: to keep "unique and/or sensitive areas . . . off limits to drilling." At a minimum, that means quashing the oil and gas industry's bid to frack inside our national forests, on the doorsteps of our national parks, or in places that provide drinking water for millions of Americans. Second, the president should call for an end to the loopholes that make fracking exempt from key provisions of our nation's environmental laws.

To the dire fracking numbers presented in our research, we should add a more hopeful one: In August 2013, Americans submitted more than one million comments urging the Obama administration to take much stronger action to protect our environment and health from fracking.

References

"Fracking by the Numbers." Environment America Research & Policy Center. October 3, 2013. http://www.environmentamericacenter.org/reports/amc/fracking-numbers. Accessed on July 19, 2014.

"Hazardous Situation." Brownfield News. July 11, 2014. http://www.brownfieldonline.com/Content/Default/Homepage-Rotating-Articles/Article/Hazardous-Situation/-3/19/2158. Accessed on July 19, 2014.

"Investigation Continues at Site of House Explosion." Fox 8 Cleveland. http://fox8.com/2014/07/17/investigation-to-continue-at-site-of-house-explosion/. Accessed on July 19, 2014.

"Oil Spill Cleanup Underway in Windsor." CBS Denver. June 21, 2014. http://denver.cbslocal.com/2014/06/21/oil-spill-cleanup-underway-in-windsor/. Accessed on July 19, 2014.

"Personal Account from the Marcellus Shale: Pam Judy." Marcellus-Shale.us. September 8, 2012. http://www.marcellus-shale.us/Pam-Judy.htm. Accessed on July 18, 2014.

John Rumpler is a senior attorney with Environment America, a federation of state-based, citizen-supported environmental advocacy organizations in 29 states (www.environmentamerica

.org). Since 2003, John has coordinated the organization's work on fracking and clean water—including state-level campaigns from California to New York. He has coauthored numerous research reports, most recently Fracking by the Numbers: Key Impacts of Dirty Drilling at the State and National Level (http://tinyurl.com/n3fsrbp). He has also testified before Congress (the House Committee on Transportation and Infrastructure) regarding enforcement of the nation's clean water laws. Prior to his position at Environment America, John practiced public interest and environmental law, representing community organizations in land use, zoning, and permitting matters. He earned his law degree at Northeastern University School of Law (1996), and studied history and philosophy at Tufts University (B.A., 1988).

Exploring Alternative Uses for Fracking Water: Lana Straub

As the boom of the oil field has increased in the early part of the twenty-first century, so has the use of water for drilling and fracking purposes. Quite a lot of the water used for oil field activities comes from nonrenewable sources of freshwater such as groundwater, particularly in the arid southwest. Nevertheless, all parts of the United States have felt the pull of water from the oil field. In the Marcellus Shale Play alone, the volume of wastewater produced from drilling activities increased significantly between 2008 and 2011, "climbing from about 365,000 m^3 in 2008 to 1,682,000 m^3 in late 2011," the increase in water usage and wastewater production corresponding directly to the upward trend in oil production (Rahm et al. 2013). Depending on the geology of the formation, "frequently millions of gallons of fracking solution are used per well" (Thornton 2014).

Concerns about proper disposal of the wastewater generated from drilling, completion, and hydrofracking processes has become an issue for local citizens as well as a national feeling of unease. "In many areas limited resources exist to properly process it. As a result, flowback solution may be improperly disposed of and potentially contaminate soil and water supplies" (Thornton 2014). To answer the concerns of tremendous water use and proper wastewater disposal, several companies are forming and research groups are completing studies to attempt to find new ways to reduce, reuse, and recycle fracking flowback water and produced water. The extraction of oil and gas can produce quite a bit of non-potable water on its own. As the deep aquifers are penetrated during the extraction process, on average, producers recover three barrels of saline-rich produce water per one barrel of oil. In some locations in Texas, the rate can be as high as 12:1 (Downey 2009).

Parts of California were hit hard by the drought conditions from 2011 to 2014. Lauren Sommer of the public radio station KQED Science desk reported in her story about Chevron providing recycled produce water to thirsty agriculture areas that claimed to have drained aquifers due to drought and oil production in the area. "In the only project of its kind in the state," reports Sommer, "Chevron's water travels several miles through a 40-inch pipe, until it arrives in a reservoir used by the Cawelo Water District." Chevron provides 26,000-acre feet and "the district mixes Chevron's water with an equal amount of freshwater, until it reaches a quality that works for local orchards" (Sommer 2014).

Moreover, Chevron is not alone. The Colorado School of Mines developed the Advanced Water Technology Center (AQWATEC) in 2006. AQWATEC's mission is "to advance the science of emerging water treatment processes and hybrid systems, enabling sustainable and energy efficient utilization of impaired water sources for potable and non-potable water supplies" (AQWATEC 2013). AQWATEC focuses on water issues in Colorado, Montana, New Mexico, Utah, and Wyoming. The group has studied potential beneficial use options of produced coal-bed methane water in these areas and found many potential use including livestock watering, agricultural irrigation, mining, fishing, wildlife maintenance enhancement, and constructed wetlands, in addition to domestic, commercial, and municipal uses.

To assist in the study of beneficial uses, AQWATEC developed a screening tool that "was used to determine treatment technology alternatives and costs, potential beneficial uses, and overall project costs for these uses. The Screening Tool also predicts treated water quality for the recommended treatment train" (Plumlee et al. 2014). Treatment trains are the types of additives or filtering systems that might need to be added to the water in order for it to provide the beneficial use. Examples of treatment trains are chemical disinfection, media filter, and tight nanofiltration (Plumlee et al. 2014).

Scientists are beginning to study how these options can be viable for fracking flowback water as well. In a study in Bear Creek, Louisiana, scientists from the Colorado School of Mines used forward osmosis (FO) as a method of reclaiming water for beneficial use. Their findings were significant. "For each well, more than 80 percent of the drilling waste can be treated, providing more than 20 percent of the water required for hydraulic fracturing of new wells." The scientists found that not only did the use of forward osmosis reduce water use, it actually made the water useful for more drilling, which reduced the need for other potable water sources. "Overall, it was demonstrated that FO is an effective technology for concentration of drilling wastewater, facilitating water reuse for fracturing operation and reducing the need for an additional water source" (Hickenbottom 2013).

One of the drawbacks to promoting beneficial use of fracking water has been the mysterious "proprietary" ingredients in the frack fluid cocktail that frighten everyday consumers. Scientists are continually conducting studies on the potential hazards of fracking fluid, which are yet to be completed; there have not been many citable examples of beneficial uses of fracking fluid besides the Chevron work in California. However, some companies are actively and successfully treating and recycling frack flowback water for reuse in the oil field, which reduces the oil field use of nonrenewable water sources. Aqua-Pure Ventures has applied the practice in several shale plays. According to company marketing literature, "[t]he Company has now recycled more than 700 million gallons of Devon Energy's flowback and produced water in the Barnett. The treated water, which would otherwise have been injected into disposal wells and permanently removed from the hydrological cycle, is instead re-used in Devon's drilling operations" (Aqua-Pure Ventures 2014).

As long as unconventional drilling continues to flourish nationally as well as internationally, flowback frack water will be an issue of discussion. New studies are ongoing on the subject of flowback fluid from hydraulic fracturing operations to determine its hazards and possible benefits to society now and in the future.

References

Aqua-Pure Ventures. "Service & Operations: Barnett Shale." http://www.aqua-pure.com/operations/shale/barnett/barnett.html. Accessed on July 31, 2014.

AQWATEC. AQWATEC Produced Water Treatment and Beneficial Use Information Center. January 1, 2013. http://aqwatec.mines.edu/produced_water/about/index.htm. Accessed on July 31, 2014.

Downey, Morgan. 2009. *Oil 101*. New York: Wooden Table Press.

Hickenbottom, Kerri L. 2013. "Forward Osmosis Treatment of Drilling Mud and Fracturing Wastewater from Oil and Gas Operations." *Desalination*. 312(3): 60–66. http://linkinghub.elsevier.com/retrieve/pii/S0011916412003086. Accessed on August 1, 2014.

Plumlee, Megan H. et al. 2014. "Coalbed Methane Produced Water Screening Tool for Treatment Technology and Beneficial Use." *Journal of Unconventional Oil and Gas Resources.* 5(3): 22–34. http://linkinghub.elsevier.com/retrieve/pii/S2213397613000372. Accessed on August 1, 2014.

Rahm, Brian G. et al. 2013. "Wastewater Management and Marcellus Shale Gas Development: Trends, Drivers, and Planning Implications." *Journal of Environmental Management.* 120: 105–113. doi:10.1016/j.jenvman.2013.02.029. http://www.ncbi.nlm.nih.gov/pubmed/23507249. Accessed on August 1, 2014.

Sommer, Lauren. "California Farmers Look to Oil Industry for Water." KQED Science. April 7, 2014. http://blogs.kqed.org/science/audio/california-farmers-look-to-oil-industry-for-water/. Accessed on July 31, 2014.

Thornton, S. L. 2014. "Fracking Solution." *Encyclopedia of Toxicology*, Volume 2. http://dx.doi.org/doi:10.1016/B978-0-12-386454-3.01229-X. Accessed on August 1, 2014.

Lana Straub is a freelance journalist and radio producer from West Texas and a regular contributor to the Water Well Journal Magazine *and KXWT-West Texas Public Radio. She is a 2014 IJNR fellow and produces her own multimedia blog, WaterTells.*

Fracking Like It's Your Job: Why Worker Safety in the Fracking Industry Is So Important: Laura Walter

Let's face it. Fracking is controversial, and that controversy can get ugly. But no matter how strongly one might oppose the process of fracking and its potential environmental impact, there is one population that always deserves to be protected: the workers in the field.

Work is an integral part of one's life. Whether the job is full-time or part-time, professional or blue collar, salaried or hourly, temporary or long-term, each and every employee has the right to a workplace that will not negatively impact his or her safety, health, and well-being. This applies to all workplaces, whether it is an office, where the biggest risk might be a paper cut, or a job in a more inherently dangerous line of work, such as construction, electrical work, or oil and gas excavation—including fracking.

Thanks to the Occupational Safety and Health Act of 1970 (OSH Act), American workers are protected under federal law from workplace health and safety hazards. Sometimes, this protection translates to safety equipment such as hard hats, fall protection, high-visibility clothing, or respiratory protection. Every safety expert worth his or her salt, however, knows that the true key to safety is not merely combating existing safety concerns with additional equipment, but rather taking those hazards out of the equation entirely. Furthermore, training employees adequately and fostering a positive workplace safety culture will do more to ensure safety than simply issuing a list of rules. The bottom line is that everyone who goes to work should be able to expect to return home healthy and safe by the end of that shift—and it is the employer's responsibility to make sure that happens.

It is true that workplace safety in the United States has improved in leaps and bounds in the decades since the OSH Act was signed. But there is still a long way to go. Thousands of employees continue to die preventable deaths on the job every year in this country. Those fatalities are represented in definable incidents, such as falls from height, traffic accidents, contact with machinery, and chemical exposure. What is often more difficult to quantify are the thousands of additional deaths each year from occupational illness. Breathe in some dust or gas on the job one year and then come down with a fatal lung disease a few years later, all linked to that on-the-job exposure—this type of occupational hazard can be tricky to track or monitor, but it happens to American workers all the time. And it appears it may be happening right now within the fracking industry.

One of the biggest risks fracking workers face is potential exposure to respirable crystalline silica. According to the Occupational Safety and Health Administration (OSHA), the federal agency tasked with protecting American worker safety, the mineral silica makes up the sand used in the hydraulic fracturing process (Silica Exposure during Hydraulic Fracturing 2014). If workers inhale this microscopic airborne dust, they are at risk of serious or fatal diseases including lung cancer or silicosis, a possibly fatal lung disease, to name just two such conditions. Both OSHA and the National Institute for Occupational Safety and Health (NIOSH) note that "large quantities of silica sand are used during hydraulic fracturing," with multiple sources of silica dust exposure inherent in the fracking process (Worker Exposure to Crystalline Silica during Hydraulic Fracturing 2012). This means that workers in the fracking industry are vulnerable to a highly dangerous airborne dust that could lead to debilitating or even fatal illnesses.

In an ideal world, all employees would be adequately protected from exposure to threats such as respirable silica. Unfortunately, we don't live and work in an ideal world. While OSHA has set permissible exposure limits for silica, NIOSH sampled the air around 116 fracking sites in 2012 and found that nearly half indicated silica exposure greater than OSHA's limits, with 9 percent indicating exposure levels at least 10 times the limit. NIOSH therefore concluded that at these fracking sites, "an inhalation health hazard existed for workers exposed to crystalline silica" (Worker Exposure to Crystalline Silica during Hydraulic Fracturing 2012).

For its part, OSHA initiated a long-awaited proposed rule for silica in late 2013. OSHA claims the ruling, if implemented, "could save nearly 700 lives and prevent 1,600 new cases of silicosis per year" (OSHA Crystalline Silica Rulemaking 2014). But this is just the beginning. In general, new federal safety regulations face long periods of scrutiny along with criticism from industry interests who might view the new rules as an impediment to economic growth, speed of operations, and profits. Business versus safety is an age-old battle, and time will tell how and when OSHA's proposed silica rule can make a difference within the fracking industry and beyond.

Silica isn't the only danger that fracking employees face. The oil and gas industry is considered one of the riskier trades for workers, with hazards including toxic chemical spills, explosions, contact with heavy machinery, transportation accidents, falls, and more. Recently, NIOSH investigated four fracking worker fatalities caused by chemical exposure. NIOSH pointed out that chemicals involved in fracking flowback operations are dangerous and "can affect the eyes, breathing, and the nervous system" (Reports of Worker Fatalities during Flowback Operations 2013).

Clearly, fracking is rife with potential risks to its employees. Just as communities need access to clean water and air, workers need access to jobs that will not endanger their health and safety. Fracking might be a controversial practice, but that does not change the fact that workers in the industry deserve adequate protection from chemical exposure, silicosis, and any other workplace hazard. After all, the importance of keeping workers safe goes deep—deeper than any fracking operation ever could.

References

"OSHA Crystalline Silica Rulemaking." Occupational Safety and Health Administration. https://www.osha.gov/silica/. Accessed on July 15, 2014.

"Reports of Worker Fatalities during Flowback Operations." NIOSH Science Blog. May 19, 2013. http://blogs.cdc.gov/niosh-science-blog/2014/05/19/flowback/. Accessed on July 15, 2014.

"Silica Exposure during Hydraulic Fracturing." OSHA Infosheet. https://www.osha.gov/dts/infosheets/silica_hydraulicfracturing.html. Accessed on July 15, 2014.

"Worker Exposure to Crystalline Silica during Hydraulic Fracturing." NIOSH Science Blog. May 23, 2012. https://www.osha.gov/dts/hazardalerts/hydraulic_frac_hazard_alert.html. Accessed on July 15, 2014.

Laura Maylene Walter is a writer and editor living in Cleveland. As the past senior editor of EHS Today, *she has written many articles about occupational and environmental safety, with a special interest in workplace culture, health, and productivity. She is the author of the award-winning short story collection* Living Arrangements, *and her writing has appeared in publications including* The Sun, Poets & Writers, Cat Fancy, *and various literary journals.*

Feeding the Fracking Workforce: How Worker Nutrition Supports Health, Safety, and Productivity: Christopher Wanjek

Remote and temporary work sites prevalent in fracking operations present a challenge for housing and feeding workers. The nearest town and accommodations might be hours away by car; often such work sites are fly-in operations. Hostile environments—extreme cold or heat—and inadequate roads and other infrastructure limitations compound the problem of providing adequate food and shelter. Inadequate accommodations can leave workers vulnerable to poor nutrition, sickness, inadequate rest and sleep, long-term health problems, low morale, and general apathy about the work at hand. This, in turn, can lead to lower productivity and increased risk of accidents. This article examines factors that affect worker health, safety, and productivity at fracking sites and provides the rationale of how better feeding and relaxation programs can increase productivity and lower accident rates.

Food as Protection

Accidents are a fracking operation's worst nightmare. One wrong switch, one loose bolt, one missed safety check, and the entire system can blow. Workers' lives are at stake. And, depending on the severity of the accident, the life of the entire community is threatened, too, as well as the very life of the company and the broader business climate for years to come. Years of a carefully cultivated corporate social responsibility practice can evaporate overnight.

Approximately 90 percent of workplace accidents are caused by human error (Feyer and Williamson 1998), and the number one cause of worker-initiated accidents is fatigue in its various forms, such as exhaustion, weakness, or sleepiness (Chan 2009). Workers need to understand not only how to operate equipment but to make quick yet prudent decisions, when necessary. Yet no amount of training can prepare a sleepy, fatigued, or otherwise unfocused or unmotivated worker from making a poor decision or a wrong move.

Companies that complement worker safety and training programs with adequate feeding and relaxation programs report lower accident rates and higher productivity and morale (Wanjek 2005). Food and rest are, in essence, protective equipment, and they serve to lower the risk of serious workplace accidents. As such, food and rest should be viewed as essential to worker safety and health as goggles, ear protection, or the sundry protective elements found at any industrial site.

What Workers Need, and What Workers Get

In many parts of the world, remote industrial sites are synonymous with poor worker accommodations and vast environmental degradation. Think mining operation and shantytowns. Such sites can belittle an international company's global reputation. Worker accommodations in the fracking industry are far better in North America and other economically developed regions, but they still are far from ideal (Wanjek 2005). In many remote-site operations, when the call for work goes out, workers must scramble for housing (Dragseth 2011). Many workers pile up in local motels, if available, with the word *local* meaning anywhere within a two-hour drive from the work site. Other workers tow trailer homes close to the work site, and these are usually ill suited for long-term use or for extreme heat or cold.

Securing food daily quickly becomes a problem. Even if these workers have access to a stove and know how to cook, they will have difficulty obtaining fresh foods, and their meals will likely be basic and nutritionally inadequate for optimal health. Those living in motels often are forced to eat at the same diner or fast-food restaurant for the duration of their contract, which could be months. Imagine such an existence, all the while working 12-hour days.

Workers at remote and temporary fracking sites are a diverse crew. They might be college-educated engineers. They might be high-school-educated blue-collar workers. They are mostly men (although women are increasing their numbers). But they often have one thing in common: They likely are leaving their family and the comforts of home for hard work and long hours. For most workers at remote sites, anything the company can do to relieve the stress of long hours and great distances will be a plus for worker health and morale and, by extension, for company safety and productivity.

Successful food and housing programs at remote work sites —where *successful* means high morale, high productivity, low absenteeism, low turnover, and few, if any, serious accidents —have just a few key ingredients: CEOs or upper management that truly care about worker safety and health; thoughtful catering that takes into consideration the nutritional needs of the workforce; opportunities to relax with coworkers; and the feeling of a home away from home, as opposed to a bunker away from home (Wanjek 2005).

Successful meal programs cater to the unique nutritional needs of a heavy laborer at a fracking site. Most workers will burn more than 300 kcal/hour. A carbohydrate gruel leaves workers tired after just a few hours; fast foods leave workers chronically tired after just a few days. In general, the body in manual labor requires 1.2–1.7 grams of protein per kilogram body weight (Lemon 1998). The B-complex vitamins, found in meats and whole grains, are needed for tissue repair and energy conversion. Vitamins A and C, found in vegetables, are needed for immunity. Vitamin D, found in fatty fish (and sunshine, if available), is needed for calcium absorption. The nutritional needs are well known and, indeed, employed by professional sports teams (Borden 2014).

Return on Investment

Investments in feeding and wellness programs routinely yield profits in terms of higher productivity, fewer accidents, and absenteeism and turnover (Wanjek 2005). Husky Injection Molding Systems, Ltd., in Bolton, Ontario, has reported a US $6.8 million yearly saving from a US$2.5 million investment in wellness. Coors Brewing Company has reported a $6.15 productivity gain for every $1 invested in food and fitness. Similarly, Travelers reports a $3.40 gain for $1 invested in nutrition and recreation; DuPont reports a $2.05 gain for $1 invested in nutrition and recreation; and the Canadian government estimates that wellness, in the form of better feeding, health promotion, and relaxation programs, amounts to a CAN$2 to $6 return on investment (Wanjek 2005).

And food and housing is just that: an investment. No company is required to provide healthy food and comfortable accommodations. Yet it is little coincidence that the best companies to work for, as ranked by *Fortune* magazine, routinely have superior cafeterias and various recreational comforts.

Remote-site food and housing accommodations, if planned well, can have numerous positive effects on profit and productivity. Tangible and immediate benefits can include well-rested and well-nourished workers with high morale and productivity levels and lower risk for serious accidents. Long-term benefits for the fracking industry from such an investment can include community support and a positive corporate image.

References

Borden, Sam. 2014. "A Squad as Intense as Its Enchiladas." *The New York Times*. June 28, 2014, SP7.

Chan, M. 2009. *Accident Risk Management in Oil and Gas Construction Projects in Mainland China*. Sydney: University of Sydney.

Dragseth, D. "Help Wanted: The North Dakota Boom." New Geography, October 26, 2011. http://www.newgeography.com/content/002501-help-wanted-the-north-dakota-boom. Accessed on July 14, 2014.

Feyer, A. M. and A. M. Williamson. 1998. "Human Factors in Accident Modelling." In: J. M. Stellman, ed. *Encyclopaedia of Occupational Health and Safety*, 4th ed. Geneva: International Labour Organization.

Lemon, P. W. R. 1998. "Effects of Exercise on Dietary Protein Requirements." *International Journal of Sport Nutrition*. 8 (4):426–447.

Wanjek, Christopher. "Food at Work: Workplace Solutions for Malnutrition, Obesity and Chronic Diseases." Geneva: International Labour Organization. 2005. http://www.ilo.org/wcmsp5/groups/public/@dgreports/@dcomm/@publ/documents/publication/wcms_publ_9221170152_en.pdf. Accessed on July 14, 2014.

Christopher Wanjek is a health writer based in Washington, D.C. He is the author of Food At Work: Workplace Solutions for Malnutrition, Obesity and Chronic Diseases, *written for the United Nations' International Labor Organization. The book documents how poor nutrition in the workforce is a health, safety, and productivity concern. Published in 2005, the book has inspired laws enhancing workers' nutrition globally.*

xt for the Marcellus and its
and
Laurentia
Depositional basin for Marcellus
389 Ma
Early Acadian
Gondwana

Introduction

One avenue toward a better understanding of the history and nature of hydraulic fracturing comes through learning about individuals and organizations who have made significant contributions in the field. It is impossible to mention every person and organization who has made such a contribution in a book of this length. However, the biographies and reviews provided here provide a taste of the way that individuals, agencies, corporations, and other institutions have made the field of hydraulic fracturing what it is today.

Many of the organizations working for and against hydraulic fracturing are large corporations, such as multinational oil and gas companies or other corporate associations, or national or international nongovernmental or other nonprofit associations. Examples of such groups on both sides of the fracking debate are listed here. But, as with many environmental issues, many of the groups that oppose fracking are relatively small, generally underfunded and understaffed, local groups organized to deal with specific problems associated with fracking, rather than

Dr. Terry Engelder, Professor of Geology at Penn State University, left, is assisted by Susan Oliver while showing former Pennsylvania Governor Tom Ridge a chart explaining the Marcellus shale formation while touring outcroppings near Williamsport, Pennsylvania on June 16, 2011. (AP Photo/Ralph Wilson) the greater debate over fracking nationwide or worldwide. It is impossible to list many such local organizations, so one has been selected to give a flavor of the type and characteristics of such action, the No Fracked Gas in Mass group.

American Gas Association

400 North Capitol St., NW
Washington, DC 20001
Phone: (202) 824-7000
URL: http://www.aga.org/
Email: none provided

The American Gas Association (AGA) was formed on June 6, 1918, with the merger of two preexisting organizations, the American Gas Institute (AGI) and the National Commercial Gas Association (NCGA). The former organization was interested primarily in technical issues, while the NCGA focused on problems of marketing natural gas. The newly formed AGA combined the mission and activities of its two predecessors.

At the time, the vast majority of gas used was called *manufactured gas*, a product made synthetically by a variety of means. There was little or no market at the time for natural gas. That situation began to change during the 1920s, however, as natural gas found a greater variety of uses than in lighting, which had been the primary application of manufactured gas. Reflecting this change in emphasis, the AGA merged with the Natural Gas Association of America (NGAA) in 1927, providing the basic institution that exists today as the American Gas Association. Today the organization claims to represent more than 200 local energy companies with more than 68 million customers in the United States. Its primary objectives are to:

- Serve as an advocate for issues related to natural gas use that are of primary concern to member companies;

- Promote the efficient use of natural gas;
- Assist members in achieving excellence in the safe, secure, efficient, reliable, and environmentally sensitive delivery of their product;
- Assist member companies in responding to customer needs;
- Collect, analyze, and disseminate information about natural gas issues of relevance to member companies; and
- Encourage the "identification, development, commercialization, demonstration and regulatory acceptance" of technologies that will allow natural gas companies to compete successfully in the marketplace.

Much of the AGA's work is carried out through more than 50 committees formed to deal with topics of special interest to the association. Examples of those are committees on building energy codes and standards, climate change, corrosion control, distribution and transmission engineering, environmental matters, environmental regulatory action, government relations policy, labor relations, natural gas security, plastic materials, safety and occupational health, state affairs, sustainable growth, and underground storage.

The current list of issues in which AGA is most interested tends to reflect the list of committees with major topics being subjects such as climate and energy policy, energy efficiency, environment, fiscal responsibility, natural gas vehicles, the promise of natural gas, rates and regulatory issues, responsible natural gas development, safety, security, supply, and taxes.

Somewhat interestingly, the AGA appears to have no special committee or issue that focuses on issues related to hydraulic fracturing, although information on the technology is available in various places on the organization's website. As one might expect, its position tends to be generally supportive of fracking as a clean and responsible method for extracting natural gas from reservoirs.

American Petroleum Institute

1220 L St., NW
Washington, DC 20005-4070
Phone: (202) 682-8000
URL: http://www.api.org/
Email: http://www.api.org/contactus.cfm

The American Petroleum Institute (API) is a trade advocacy association consisting of more than 400 companies involved in the production, refining, and distribution of petroleum, as well as support functions for the industry. The association was founded on March 20, 1919, after the end of World War I. During the war, a number of independent petroleum companies had worked together and with the federal government to produce and supply the petroleum products needed to win the war. When the war was over, the petroleum companies recognized the benefits of having a single trade organization that could represent their interests before the government and the general public. Such an association also had the potential for establishing, promulgating, and enforcing industry-wide standards for the production and distribution of petroleum and petroleum products. The first of those standards was announced in 1925. Today, the API administers a complex and comprehensive set of more than 500 such standards covering all aspects of the petroleum and natural gas industry. The API currently has State Petroleum Councils in 33 states, all of which are located east of the Rocky Mountains.

API's activities can be broadly classified into one of four fields: standards, education and certification, statistics, and public advocacy and lobbying. Consistent with its early focus, the organization produces new and revised standards on a regular basis, covering virtually every conceivable aspect of petroleum and natural gas production and distribution. These standards cover topics such as the quality and uses of motor oil; dimensions of rings, plugs, threads, and other devices used in the industry; design and manufacture of vessels; measuring instruments and methods of calibration; safety and fire regulations; and inspection standards and procedures for refining operations. Certain API standards are well known both in and outside the industry as, for example, the "API gravity" number, which identifies the density of a petroleum product, and the "API number," which is a unique identifier for every well and exploration site in the United States.

The core of the API's training and certification effort includes programs such as: the API Monogram Program, which allows companies to apply for certification for the equipment they produce; APIQR, which allows companies to have their products certified as approved for use in the industry by the API; Individual Certification Program (ICP), which provides individuals with training in and certification for certain types of specific skills used in the industry; Witnessing Programs, which trains individuals in the skills required to observe and approve a variety of procedures used in the industry; and Energy Risk Management, in which individuals are trained to assess and deal with the variety of risks that arise during the exploration for and production and distribution of petroleum and petroleum products. In addition to the training of workers in the field of petroleum and natural gas, API produces a variety of materials for the education of the general public. One of the best known of these resources is a website called Classroom Energy (http://www.classroom-energy.org/), at which the organization presents its own view on a number of energy-related issues, such as climate change, environmental protection, and oil spills. Critics have noted that these presentations are somewhat biased toward the view of the industry, which is hardly surprising or unreasonable given that advocacy is a legitimate function of the organization and its materials.

The API is particularly and justifiably proud of its work in producing statistics about every aspect of the petroleum and natural gas industries. It began this activity very early in its history, releasing weekly data on the production of crude oil as early as 1920. Among the core data sets produced by the organization are the Weekly Statistical Bulletin, Monthly Statistical Report, Quarterly Well Completion Report, Imports and Exports of Crude Oil and Petroleum Products, Joint Association Survey on Drilling Costs, and Inventories of Natural Gas Liquids and Liquefied Refinery Gases.

Finally, the API is a strong advocate for certain industry-favorable positions on a variety of energy, environment, political, and other issues. It exercises this function in a variety of ways, such as television advertising and news releases to the media. It also expresses its views through a number of lobbyists (16 in 2009) that attempt to influence legislation on issues related to petroleum and natural gas. In 2009, its budget for public advocacy was reported to be $3.6 million. Some of its positions on important national issues include support for additional oil and gas drilling as a way to increase the number of jobs available for Americans and improve the overall economy, for an extension of tax benefits to the oil and gas industry as a way of encouraging further exploration and improving the industry's efficiency, and for voluntary industry regulation as a way of dealing with environmental issues related to oil and gas production and distribution.

As do most trade organization, the API sponsors and participates in a number of meetings, conferences, workshops, and other sessions in which new information can be disseminated and participants can interact with each other. Some of the topics of such meetings are exploration, production, distribution, offshore wells, storage tanks, oil and gas tankers, facility security, regulations, standards, and training and certification. The API also sponsors a prodigious publication program, producing more than 200,000 publications each year. Its online publications catalog includes sections on exploration and production, health and environmental issues, marine issues, marketing, refining, safety and fire issues, storage tanks, and petroleum measurement standards.

America's Natural Gas Alliance

701 8th Street NW, Suite 800
Washington, DC 20001
Phone: (202) 789-2642
URL: http://anga.us/
Email: info@anga.us

America's Natural Gas Alliance (ANGA) provides relatively little information about its own history. But outside sources report that the organization was formed in March 2009 primarily for the purpose of lobbying the U.S. Congress as it considered legislation for dealing with global climate change. A consortium of some of the nation's and world's largest natural gas producing companies, ANGA has lobbied for the presumption that natural gas should and can be considered a fuel of the future because of its relatively low input to climate change, compared to other fossil fuels. The organization currently has 21 members, including Apache, Cabot Oil and Gas Corporation, Chesapeake Energy, Energen, Laredo Petroleum, Pioneer Natural Resources, Southwestern Energy, and Xto Energy. President and CEO of the organization is Marty Durbin, nephew of the senior senator from Illinois, Dick Durbin, and formerly vice president of the American Petroleum Institute.

ANGA's declared mission is "to promote growing demand for and use of our nation's vast domestic natural gas resources for a cleaner and more secure energy future" (Statement at Quadrennial Energy Review Public Meeting; http://www.anga.us/media/press/27734331-5056-9F69-D4B0095359F599FC/files/4.11.2014%20ANGA%20QER%20Written%20Statement.pdf. Accessed on July 14, 2014). The organization has identified five major areas of interest to which it devotes its attention and energies: power generation, transportation, save and responsible development, liquid natural gas (LNG) exports, and testimony and filings. With regard to the issue of power generation, for example, ANGA expresses interest in topics such as the role natural gas can play in the nation's energy equation, the infrastructure needed for delivering natural gas to consumers, problems and issues relating to the cost of natural gas for ratepayers, and the benefits of using natural gas as a source of clean energy. The issue of save and responsible development includes an extended discussion of hydraulic fracturing and directional drilling. The web page on this issue provides a general introduction to the practice of fracking and reviews some of the questions that have been raised about the practice. It concludes with the observation that "[h]ydraulic fracturing is routinely and safely used in communities across the country."

This statement, perhaps somewhat oversimplified, reflects the fact that ANGA is, quite naturally, an enthusiastic supporter of hydraulic fracturing in the exploration for and extraction of natural gas. It is, however, somewhat difficult to find information about the topic of the ANGA website. The Media Room of the website does contain a number of news stories and opinion pieces from the organization and from other sources, although they can be found by only specific searches for the term *fracking* and related terms.

The ANGA website also includes a blog where interested parties can exchange ideas on important issues related to natural gas. The authors of the blog, other than ANGA itself, do not appear to be identified, and the blogs rarely, if ever, actually contain any exchange of ideas or opinions. Some of the blog topics are natural gas vehicles, preservation of the Gulf of Mexico habitat during resource extraction, national policies on LNG permitting, and natural gas exploration and extraction in specific localities in Arkansas, Colorado, Pennsylvania, and other states.

Chesapeake Climate Action Network

6930 Carroll Ave, Suite 720
Takoma Park, MD 20912

P.O. Box 11138
Takoma Park, MD 20913
Phone: (240) 396-1981
Fax: (888) 428-3554
Virginia Office & Mailing Address
1108 E. Main St., Suite 603
Richmond, VA 23219
URL: http://chesapeakeclimate.org/
Email: http://chesapeakeclimate.org/contact-us/

Chesapeake Climate Action Network (CCAN) was founded in 2002 by Mike Tidwell, author of a number of books about environmental issues including *Bayou Farewell: The Rich Life and Tragic Death of Louisiana's Cajun Coast* (Vintage, 2007), *The Ponds of Kalambayi: A Peace Corps Memoir* (Lyons Press, 2011), *Amazon Stranger: A Rainforest Chief Battles Big Oil* (Lyons Press, 1996), *In the Mountains of Heaven: True Tales of Adventure on Six Continents* (Lyons Press, 2003), and *The Ravaging Tide: Strange Weather, Future Katrinas, and the Coming Death of America's Coastal Cities* (Free Press, 2006). By the early 2000s, Tidwell had become interested in issues of global climate change and had converted his own home in Maryland to a nearly entirely non-fossil-fuel operating system. In 2001, he convinced the Rockefeller Brothers Fund to make a grant for the support of a new organization focusing on climate change issues in the Maryland/Virginia/Washington DC area, an organization that was to become CCAN.

The organization's mission is to build and mobilize a grass-roots movement in the region around the nation's capital that will begin to respond to the threats posed by global warming by developing policies and practices, which will lead to climate stability. It carries out its mission through a number of specific campaigns centered in each of its three geographical areas: Maryland, Virginia, and the District of Columbia. Examples of its Maryland campaigns include the Stop Cove Point: No Fracked Gas Exports; Double Maryland's Clean Power:

40% by 2025; Protect Marylanders from Fracking; Accelerate Maryland's Climate Action Plan; and Fight Dirty Coal and Incineration. Its Virginia campaigns include Safe Coast Virginia; Climate Solutions, Not More Pollution; and No Offshore Drilling for Virginia. The DC and federal campaigns focus on the District's renewable energy law, the proposed Keystone XL pipeline, coal exports, and cap and dividend approaches to the control of climate change. CCAN also sponsors comparable types of campaigns at college campuses in Maryland and Virginia.

CCAN promotes its ties with a large number of organizations in the Maryland/Virginia/DC region with similar interests, such as the Center for Health, Environment, and Justice; Green America; the Hip Hop Caucus; Physicians for Social Responsibility; Environment Maryland; Food & Water Watch; Waterkeeper Alliance; Southern Appalachian Mountain Stewards; Southern Environmental Law Center; and DC Environmental Network.

CCAN has taken a particularly active stance in opposition to fracking because of its concerns about the release of the greenhouse gas methane at a number of points in the fracking process. One of its most recent campaigns has been aimed at plans to build a large pipeline from the oil fields of Pennsylvania to a shipping terminal at Cove Point, Maryland, that was originally build to receive imported natural gas. Current plans call for converting that terminal for the exportation of LNG, primarily to Asia. Critics such as CCAN have argued that the new terminal would cause a number of environmental and other problems, such as compromising the ecology of Chesapeake Bay, increasing the accidental release of methane to the atmosphere, increasing the cost of natural gas in the United States, and forcing property owners to give up a portion of their land over which the proposed pipeline is to run. CCAN also operates a second campaign focusing on what it sees as the risk that Maryland will become the next step to allow hydraulic fracturing, given the number of gas basins that underlie the state.

Consumer Energy Alliance

2211 Norfolk, Suite 410
Houston, TX 77098
Phone: (713) 337-8800
URL: http://consumerenergyalliance.org/
Email: info@consumerenergyalliance.org

The Consumer Energy Alliance (CEA) calls itself "the voice of the energy consumer." The organization was founded in 2005 with the expressed purpose of "build[ing] a dialogue between energy providers (oil, gas, wind, solar, nuclear, etc.) and energy consumers (industry users, like chemical companies, agriculture & transportation, as well as average citizens)." ("Maximizing Your Energy Dollar," Coveritlive, http://www.coveritlive.com/index2.php/option=com_altcaster/task=viewaltcast/template=/altcast_code=c19f518d2a/ipod=y. Accessed on July 15, 2014.) The extent to which CEA is actually a grassroots consumer organization, as it claims to be, has been questioned by some individuals and organizations. A review of the organization's activities on Salon.com, for example, suggests that CEA is a "front" organization for Big Oil and Big Gas, supported primarily by multinational petroleum companies such BP, Chevron, ExxonMobil, Marathon, Shell, and Statoil. A representative of the Natural Resources Defense Council has called CEA a "front group that represents the interests of the oil industry" ("Big Oil and Canada Thwarted U.S. Carbon Standards," Salon, http://www.salon.com/2011/12/15/big_oil_and_canada_thwarted_u_s_carbon_standards/. Accessed on July 15, 2014).

CEA characterizes itself in a very different way, calling itself "the voice of the energy consumer." It says that its main purpose is to "provide consumers with sound, unbiased information on U.S. and global energy issues." More detailed information about the CEA and its activities is available on the organization's website and, in particular, in its annual reports, which are also available on the website.

CEA claims a very large membership of more than 230 organizations from the fields of academia (6 members); consumers, business, agriculture, and industry (144 members); and energy providers and suppliers (86 members). As an example, the range of members includes the American Bus Association, Beaver County (Pennsylvania) Chamber of Commerce, Colorado Farm Bureau, Elite Parking Services of America, Florida Chamber of Commerce, Marine Engineers Beneficial Association, National Association of Home Builders, Southeastern Fisheries Association, Houston Museum of Natural Science, University of Texas Center for Energy Economics, Alaska Miners Association, BP, Entergy, Ohio Oil & Gas Association, Shell Oil, Shell Wind Energy, and the U.S. Department of Energy Gulf Coast Clean Energy Application Center.

The organization works with and through 6 regional chapters that cover 20 states, CEA-Alaska, CEA-Florida, CEA-Texas, CEA-mid-Atlantic, CEA-Midwest, and CEA-Southeast. The CEA website provides news and information about events that are occurring within and as a result of the work in each of these regional chapters.

Activities of the CEA fall within three general categories, which the organization calls Educate, Advocate, and Act. The first category refers to the educational efforts of the organization to provide consumers with information about a variety of energy options, such as coal, nuclear, and wind. The description section of the organization's website provides links to additional sources of information and home websites for groups interested in each of these areas. The Advocate arm of CEA's efforts focuses on efforts made by the organization to put forward certain positions on a variety of energy issues, such as construction of the Keystone XL pipeline, trucking safety, and fracking operations in oil and gas operations. A particular feature of this segment of CEA activities is a program called Energy Voices, in which individuals and companies describe their experiences with specific issues of interest to CEA and its followers. The Act arm of the CEA program invites consumers to express their own views on a variety of issues of interest to them and to CEA, focusing on the opportunity for consumers to send letters to legislators and other policymakers about topics such as the Keystone pipeline and fracking operations.

One of the most useful sections of the CEA website is the Latest News page, which provides interesting and valuable links to a host of stories about current events in the field of various fields of energy.

H. John Eastman (1905–1985)

Eastman was associated with oil and gas drilling for most of his adult life. He is best known today for his discovery of a method for determining the direction in which a well was being drilled. When that direction was entirely vertical, such a technology was largely unnecessary. But as petroleum engineers began to consider the possibility of drilling in some direction other than true vertical—so-called directional drilling—it became essential to know precisely where a drill hole was headed in order for it to reach the predetermined oil or gas reservoir for which it was intended.

Historically, a rather crude technology called the acid-bottle test was used for this purpose. In this test, a glass bottle containing strong acid was lowered into a drill hole. When the bottle was left in position for a period of time, it began to etch the glass. The etching pattern showed the angle at which the surface of the acid was lying in the bottle. This angle could then be used to calculate the angle of the drill hole to true vertical. The test was reasonably accurate, but it took a fair amount of time, and the acid bottle had to be lowered and raised each time a reading was needed.

Eastman invented a much simpler and more efficient method for taking such measurements. In 1929, Eastman invented a device consisting of a long metal tube with the diameter of a milk bottle in which were suspended a small camera, a plumb bob, and a compass. The camera allowed a person on the surface to take pictures of the compass and the plumb bob to determine the orientation of the drill hole as it was being drilled. With this information, an operator could adjust the direction of a drilling operation as he or she wished, making possible for the first time a technique now known as *directional drilling*, of which the widely popular *horizontal drilling* is one current example.

Eastman also took his research one step further when he invented a system for changing the direction of a drill bit as it traveled through a drill hole. The key element in Eastman's system was a device known as a *whipstock*, a piece of metal pipe that changed the direction of the drill in the process of burrowing into rock. With his camera system and whipstock, Eastman was able to both follow the path of the drilling and to control the direction in which he wanted the drill to move. In commenting on Eastman's inventions, a May 1934 article in *Popular Science Monthly* magazine observed that "only a handful of men in the world have the strange power to make a bit, rotating a mile below ground at the end of a steel drill pipe, snake its way in a curve or around a dog-leg angle, to reach a desired objective." For these accomplishments, Eastman is often referred to as the father of directional drilling.

H. (Harlan) John Eastman was born in Cedar Rapids, Iowa, on July 29, 1894. At the age of seven, he moved with his family to Oklahoma City, where his father had been appointed postmaster under the administration of President William Howard Taft. After graduating from high school, Eastman matriculated at Oklahoma A&M College, now Oklahoma State University, but he did not graduate from the school.

Instead, he decided to look for work in the petroleum industry and was hired by the Magnolia Petroleum Company as its first employee. After a series of mergers and consolidations, Magnolia emerged as the Mobil Oil Company, which continues to retain Magnolia's "flying red horse" symbol.

After a short stay with Magnolia, Eastman decided to seek his fortune in California, which was undergoing an "oil boom" in the early twentieth century. There he became particularly interested in questions as to determining the direction in which a drill line was headed during drill and how that drill line might be altered for one reason or another. In working on these questions, he developed both his camera/plumb bob/compass directional system (later called a single shot drilling system) and his whipstock aiming device. He often signed on with an oil company to test and refine these devices in exchange for whatever services he could provide to the company's drilling operations. Most historians mention 1929 as the year in which Eastman formally introduced the use of directional drilling to extract crude oil at a facility in Huntington Beach, California.

Eastman achieved his greatest fame in 1933 when a huge fire broke out in the Conroe, Texas, oil field that threatened to destroy the field's petroleum assets. The field's owners invited Eastman and his colleague George Failing, of Enid, Oklahoma, to find a way of bringing the fire under control. They did so by drilling at angles into the gas and oil reservoirs at Conroe to drain away the fuels that were feeding the giant conflagration. Their success in the operation made their fame and allowed Eastman to establish his own company, Eastman Whipstock, which by 1973 had become the world's largest directional drilling company. The company merged with Norton Christensen in 1986 to form Eastman Christensen, which was sold to Baker Hughes in 1990. It continues to exist today as one of the world's largest drilling companies, Hughes Christensen, with primary office in Houston, Texas.

Eastman died at his home in Los Angeles on February 24, 1985.

Terry Engelder

Engelder is professor of geosciences at Pennsylvania State University (Penn State), where his special areas of interest are the fracture toughness of rocks and the microfabric of crystalline rocks. In previous years, he has also conducted research on the mechanical properties of rocks under strain, the fluid transport properties of permeable rocks, and the geochemistry of rock–water interactions. As well known and successful as Engelder has been in his own field, he has achieved even greater fame and, perhaps, notoriety, for his views of hydraulic fracturing and the benefits and risks it poses for modern society. He has spoken and written widely on the potential value of hydraulic fracturing as a technology essential for the recovery of natural gas resources and for his outspoken positions has earned the opprobrium of many professional colleagues as well as opponents of fracking technology. He has even acknowledged that many of his critics think of him as "the Doctor Strangelove of the fracking debate," comparing him to a somewhat demonic character in the 1964 Stanley Kubrick–directed film *Dr. Strangelove, Or: How I Learned to Stop Worrying and Love the Bomb.*

Professionally, Engelder is perhaps best known for a paper he and colleague Gary G. Lash wrote in 2008 suggesting an estimate for the amount of natural gas available in the Marcellus Play in the northeastern United States. The paper, probably for the first time, alerted the petroleum industry and the general public as to the extraordinarily large quantity of the fuel available under very large parts of a half dozen states. With that announcement, the battle over the use of hydraulic fracturing in the recovery of that resource was underway in earnest. Engelder has since acknowledge that the use of fracking in gas recovery carries with it a number of very real risks to the natural environment and human health, but that the benefits that accrue from fracking far exceed these potential hazards.

Engelder has earned his BS, MS, and PhD in geology from Penn State, Yale University, and Texas A&M University, respectively, in 1968, 1972, and 1973. He was also employed in the gas and oil industry during his years of study, first as a geologist for the Bradley Producing Company, in 1965; then as a hydrologist for the U.S. Geological Survey from 1966 to 1967; as a geologist with Texaco, in 1968; and as a research assistant at the Center for Tectonophysics at Texas A&M from 1970 to 1973. Upon graduation, Engelder accepted an appointment at Columbia University, in New York City, where he served as a research scientist, research associate, senior research scientist, and senior research associate at the Lamont-Doherty Geological Observatory, and lecturer at Columbia from 1973 to 1985. In 1985 he accepted an appointment at Penn State, where he has served as an associate professor of geosciences (1985–1990) and professor of geosciences (1990–present). Engelder has also served as Fulbright Senior Fellow at Macquarie University, Sydney, Australia (1984); French–American Foundation Fellow in France (2001–2002); guest professor at the Erherzog Johann Technical University, Graz, Austria (1999) and Università di Perugia, Perugia, Italy (2004); and visiting scholar at TotalFinaElf, CSTJF Pau, France (2001–2002).

Engelder has authored and coauthored more than 90 scholarly papers, written and cowritten about 30 book chapters, and authored one book, *Stress Regimes in the Lithosphere*. He has been honored with membership in the Phi Eta Sigma Honor Society and the Phi Kappa Phil Honor Society, election as a fellow in the Geological Society of America, awarded the Wilson Distinguished Teaching Award of Penn State, and named John and Cynthia Oualline Lecturer in Geological Sciences at the University of Texas.

Environment America

294 Washington St, Suite 500
Boston, MA 02108

Phone (617) 747-4449
URL: http://www.environmentamerica.org/
Email: http://www.environmentamerica.org/contact

Environment America was formed in November 2007 when the Public Interest Research Groups (PIRGs) decided to split off their environmental activities into a new and separate organization, Environment America. The new organization consisted of 23 state environmental organizations that had themselves separated from PIRGs earlier in that same year. As a result of the split, the PIRGs continue to purse their work in the field of consumer and social justice, while all of the organization's original environmental activities were moved to the new Environment America. A sister organization, Environment America Research & Policy Center, are respectively 501(c)(4) and 501(c)(3) nonprofit organizations. As of 2014, Environment America maintains relationships with state chapters in 29 states.

Environment America is largely an organization with a two-pronged approach: research and education. It carries out studies of important environmental issues facing the American people and then informs legislators, opinion-makers, and the general public of the results of its studies. For example, the 2013 study "Fracking by the Numbers: Key Impacts of Dirty Drilling at the State and National Level" collected data and statistics on the health and environmental impact of hydraulic fracturing deep-well injection of wastewater throughout the United States. The study brings together information about these effects that is generally not easily available from other sources.

Information obtained from research is then distributed by Environment America through a variety of means, such as interviews with newspaper, radio, and television reporters; news conferences on specific topics; op-ed pieces; letters to editors; communications with legislators and legislative staffs; and other mechanisms (such as the article by John Rumpler in Chapter 3 of this book). As of 2014, Environment America had selected eight major topics on which to focus their attention: Wind Power for America; Stop Fracking Our Future; Repower America; Protect America's Waters; Let's Get Off Oil; Global Warming Solutions; Conservation America; and Clean Air, Healthy Families. As an example, the fracking program was being carried out in cooperation with 18 state affiliates from California to Rhode Island and Minnesota to Florida. Each state develops its own program in cooperation with the national organization to bring to the people of the state information about fracking and the actions that individuals can take to deal with the problem.

One of the important functions played by the national organization is the collection, assembly, and distribution of current news about a particular issue. A recent Environment America website, for example, contained stories about legislative efforts to adopt a fracking moratorium in Massachusetts, a letter campaign among members informing President Barack Obama about the need for an end to fracking, a new EPA rule concerning the use of diesel oil in fracking operations, a collection of personal stories about the effects of fracking on people's lives, and the Pennsylvania supreme court's ruling on local zoning laws on fracking operations.

Food & Water Watch

1616 P Street NW, Suite 300
Washington, DC 20036
Phone: (202) 683-2500
Fax: (202) 683-2501
State offices in California, Colorado, Florida, Illinois, Iowa, Maine, Maryland, Michigan, New Jersey, New Mexico, New York, North Carolina, Ohio, Oregon, Pennsylvania, and the European Union (Brussels)
URL: http://www.foodandwaterwatch.org/
Email: http://www.foodandwaterwatch.org/about/contactus/

Food & Water Watch was established in 2005 when 12 members of the Energy and Environment Program of the nonprofit organization Public Citizen decided to form their own association to focus specifically on food and water issues. The current executive director of Food & Water Watch, Wenonah Hauter, was previously director of the Energy and Environment Program at Public Citizen. The current Food & Water Watch staff is divided into groups responsible for operations, communications, food, water, common resources, international, organizing, legal advocacy, development, finance and human resources, and information technology.

The issues on which Food & Water Watch is currently working are divided into three major categories: food, water, and food and water justice. Some of the specific issues in the area of food include antibiotics, consumer labels, factory farms, food safety, genetically modified foods, irradiation, nanotechnology, and radiation impacts. In the general area of water, the organization's specific campaigns focus on bottled water, desalination, fracking, radiation impacts, water privatization, world water, and a specific program called Renew America's Water.

Food & Water Watch's campaign on fracking is based on the assumption that the operation is a water-intensive activity that not only depletes a valuable resource—water—but also releases into the environment a variety of chemicals known to harm human health. The organization argues that efforts to find appropriate ways of regulating fracking are misguided and that efforts should focus instead on outright bans of fracking. The organization maintains a web page that lists more than 400 local communities that have taken action in some form or another to ban fracking from within their borders (http://www.foodandwaterwatch.org/water/fracking/fracking-action-center/local-action-documents/).

Food & Water Watch provides a host of resources to support the case in opposition to fracking, including a number of fact sheets, reports, and online web pages dealing with specific aspects of the topic. Fact sheets are available on subjects such as how the proposed Trans-Pacific Partnership will affect local communities, the involvement of college and university researchers in promoting fracking, and problems with natural gas pipelines. Some reports on fracking that are available deal with the social costs of fracking, efforts to ban fracking in the state of Maryland, why fracking and natural gas will not lead to energy independence in the United States, and New York state policies on fracking and their effect on employment opportunities in the state.

Some of the resources available on the Food & Water Watch website deal with topics such as the relationship between free trade agreements and the expansion of fracking, the relationship of fracking and global climate change, the effects of fracking operations on food production, and the ways in which fracking contributes to the global water crisis.

The Food & Water Watch website also provides a variety of multimedia presentations that focus on specific topics of interest, such as a video on the Global Grocer (where foods come from), the Meatrix (factory farming), and a number of films to rent that includes two classic movies on fracking, *Gasland* and *Gasland Part II*. The organization also conducts research on a number of topics related to its fields of interest. Reports on this research are available at no charge through the organization's website. Some examples are the industrialization of fracking sand, threats posed by wastewater from fracking for the Great Lakes, illusory benefits of fracking in the state of Colorado, the potential threats of fracking on New York state agriculture, the dangers posed by fracking on New York City's water system, and false promises of jobs offered by the fracking industry.

The Food & Water Watch website is also a valuable research for news and opinion on current developments in the field. In addition to a news and blog page, the website offers new films and videos, press releases, and a list of forthcoming events throughout the United States and Europe.

Ground Water Protection Council

13308 N. MacArthur
Oklahoma City, OK 73142
Phone: (405) 516-4972
URL: http://www.gwpc.org
Email: http://www.gwpc.org/contact-us

The Ground Water Protection Council (GWPC) was incorporated as a 501(c)(6) nonprofit corporation in the state of Oklahoma in 1983. It was originally founded by representatives from five states with the objective of dealing with underground injection control issues as well as general ground water protection questions. The organization now includes high level ground water officials from all 50 states. The organization is run by a board of directors of 20 senior state water officials who are elected by the general membership from each of the 10 U.S. Environmental Protection Agency districts in the United States. The board itself is divided into a number of committees that deal with specific issues, such as carbon dioxide geosequestration, source water, technical issues, and administration.

GWPC activities are organized primarily into three major "division": water and energy, water availability and sustainability, and water quality. Each division is chaired by a member of the board of directors and is designed to give members an opportunity to present ideas, discuss issues, and develop responses to critical issues within each division. In turn, each division selects specific topics on which to work in more detail. The water and energy division, for example, subdivides its work into working groups on hydraulic fracturing, energy-related injection, water usage, and abandoned mines. Perhaps its most important accomplishment thus far with regard to the fracking issue has been the development of a web page called FracFocus. That web page provides an interactive that allows a person to access information about any one of more than 77,000 individual oil and gas wells in the United States. By listing the identifying characteristics of a well, a person can find out which chemicals are used in the drilling and injection processes at that well. Registration on FracFocus is required in some states, and relevant information about wells is also provided voluntarily by a number of petroleum companies.

FracFocus also provides information about other aspects of the fracking process, such as the technology involved in the process, chemicals used in fracking, effects of fracking on ground water supplies, and state regulations on fracking.

Additional information about fracking is available from the energy injection working group, which studies the effects of wastewater injection from fracking processes on ground water resources.

GWPC produces a number of publications containing very useful information on a variety of topics related to ground water protection, especially in the area of hydraulic fracturing. Examples of those publications include a general introduction on shale gas; a brochure on injection wells, their use, operation, and regulation; a comprehensive and extensive report on the status of ground water resources in the United States, "Ground Water Report to the Nation . . . A Call to Action"; a white paper on the relationship between injection wells and seismicity; and papers in its Spotlight Series on topics related to fracking and its effects on ground water, such as injection wells and seismicity.

In addition to such publication, GWPC produces a monthly newsletter, "Groundwater Communiqué"; reports on the organization's conferences, forums, and other meetings; webinars on topics such as funding available from the Clean Water Act and stormwater management and source water protection; and links to dozens of governmental, nongovernmental, industry, and environmental organizations in the United States and around the world. Of special interest to both professionals in the field and the average system is the GWPA web page on News and Updates, which includes articles on a large number of specific topics, such as hydraulic fracturing, injection wells, well owners, drilling and operation regulations, and legislative actions on ground water issues.

Erle P. Halliburton (1892–1957)

Halliburton was the founder of the energy company that now carries his name and that is probably one of the most famous multinational companies in the world. It is not only an enormously successful and profitable corporation but also the subject of extensive criticism by some individuals and organizations for its business practices. The company has 13 major product lines that include drill bits and services, wireline and perforating, completion tools, cementing, testing and subsea, production enhancement, and consulting and project managements. In 1947, Halliburton performed the first trial fracking experiment for the Stanolind Oil Company in Kansas. In 2014, the company reported having about 80,000 employees of 140 different nationalities in 80 countries.

Erle Palmer Halliburton was born on a farm near Henning, Tennessee, on September 22, 1892. He attended school in nearby Ripley and had originally planned to matriculate at a college after graduation from high school. He father's death when Erle was 14 years old changed those plans, however, and he instead left school to begin work with railroad construction companies. At the age of 16 he had already learned the skills needed to operate steam cranes on the Mississippi River, a job at which he worked for two years. In 1910, he enlisted in the U.S. Navy where he was assigned to work on the Navy's first motor barge.

After being discharged from the Navy at the age of 23 in 1915, Halliburton found work with the Dominuiz Corporation in California, where he was appointed superintendent of water distribution. Very shortly thereafter he took a position with the Perkins Oil Well Cementing Corporation, a company that, as its name suggests, performed cementing operations in the installation of oil drilling rigs. He did not last long with the company primarily because of his habit of criticizing existing procedures and suggesting more efficient ways of doing things. After only about a year with Perkins, Halliburton was fired for annoying his supervisors with his ideas. He later told a biographer that the two greatest things that happened in his life were to have been hired and fired by Perkins.

Although he left the Perkins company, Halliburton did not abandon his interest in oil well cementing. He continued to work on his own to develop and improve cementing techniques, even as he had little or no money or resources with which to try out his ideas. (The story is told that he was so poor during this period that he washed out the cement bags he had used and sold them back to the cement company to make a few pennies extra.) Finally in 1920, Halliburton had his opportunity. He was hired by the Skelly Oil Company to bring under control a new well that was spewing crude oil into the air. He was able to put into practice many of the ideas about well cementing that he had been trying out, and Skelly was sufficiently impressed to offer him a full-time job. Halliburton remained with Skelly for four years until he founded his own business, the Halliburton Oil Well Cementing Company, in 1924.

Halliburton's new business grew slowly but surely. When it began operations, its only asset was one mule-drawn wagon, but three decades later, Halliburton had bought up all of his competitors and had cemented more than 100,000 oil wells in the U.S. west. He had improved his own financial standing from a point at which he was earning only $260 a month as the founder of his new company to estimated assets of about $100 million at his death in 1957. Halliburton continued his research on cementing and related drilling techniques through much of his life, eventually receiving at total of 38 patents for various oil well tools and technologies as well as unrelated topics. For example, he designed and built a line of metal suitcases that is still being made and sold by the Zero Halliburton company. In addition to his interests in petroleum exploration and extraction, he followed other lines of business that included a gold mine and a hydroelectric plant in Honduras, as well as farming and ranching activities in the United States. Late in his life be became interested in supporting a wide variety of charitable activities and organizations. He died in Los Angeles on October 13, 1957. After his death, Halliburton was inducted in the Oklahoma Hall of Fame, the Permian Basin Petroleum Museum Hall of Fame, and the Oklahoma Aviation Hall of Fame. The last of these honors recognized Halliburton's founding and operation of the Southwest Air Fast Express airline (also known as Safeway Airlines) in 1928, a company that eventually merged with American Airlines. During its brief period of operating as an independent airline, Southwest Air Fast Express offered service between St. Louis and Dallas and, later, between those cities and New York City and Los Angeles.

William Hart (1797–1865)

Historians often date the origin of the natural gas industry in the United States to the discovery of oil at Titusville, Pennsylvania, by "Colonel" Edwin Drake in 1859. While it is true that Drake's well produced gas as well as oil, this attribution is not correct. Indeed, natural gas had been discovered and put to use more than three decades earlier in an area not so distant from Titusville: Fredonia, New York. The person responsible for that accomplishment was a tinsmith by the name of William Hart.

William Aaron Hart was born in the town of Bark Hasted, Connecticut, in 1797. Essentially nothing is known about his early life, except that he moved to Fredonia in 1819, reputedly bringing with him nothing other than his rifle and a pack of clothing. Again, little is known about his time in Fredonia until about 1825, except that he apparently pursued his craft as a tinsmith in the town. At some point, he became interested in the possibility of capturing and putting to use a seep of natural gas present adjacent to Canadaway Creek running through the center of town. His approach was to dig a hole (using shovels!) that eventually reached a depth of 27 feet to gain improved access to the natural gas. He then trapped the gas in a "gasometer," originally his wife's laundry tube with a hole cut into the bottom. The gasometer provided not only a reservoir for the gas, but also a means by which it could be measured (hence the term *gasometer*). Gas from the well was then transported to nearby homes through a series of hollow wooden pipes whose junctions were sealed with pitch. By August of 1825, Hart's natural gas system was being used to heat two stores, two shops, and one mill in the vicinity of the well. Three months later, 36 gas lamps had been installed in the town. Hart's project had more than utilitarian value as residents and visitors to the town, who came from "far and wide," were astounded at the spectacularly beautiful, as well as useful, scene produced by the combustion of pure natural gas from the ground.

Hart soon expanded his efforts, extending the original well to a depth of 50 feet in 1850. At that point, he was producing enough gas to light 200 lamps in the area. Realizing the potential commercial value of Hart's operation, a group of Fredonia residents incorporated the Fredonia Gas Light Company in 1828, the first commercial natural gas company (and now the oldest) in the United States. The company continued to expand and improve its natural gas system in 1858 by digging a second well to a depth of more than 200 feet. By this time, the method of delivering natural gas to customers had also been vastly improved. The original wood-and-pitch system was, as one might guess, woefully inadequate because of the ease with which gas escaped from the pipes. The system was eventually replaced by a new method of delivery consisting of lead pipes that carried gas from the well to a store, shop, mill, or other location. The gas was then distributed throughout the site by means of finer pipes made of tin (of course, Hart's specialty), which delivered a very small flow of gas to a room equivalent in brightness to about two normal candles. Turning a gas lamp on and off was the simplest of operations. During the day, a plug was placed in the end of a tin pipe, and at night it was removed and the gas lighted.

The rest of William Hart's life is nearly as much of a mystery as it earlier years. He established a home at 50 Forest Place in Fredonia, where he abandoned his work with natural gas and became a nursery operator, best known for the roses he grew. He also operated an amusement park that included a hot-water spa that, some historians believe, may have been heated by the combustion of natural gas. At some point, Hart and his family moved to Buffalo, where census records later listed his occupation as "merchant," and that of his adult son as "gas furnisher." Hart died in Buffalo on August 9, 1865, leaving behind his wife, son, and one daughter.

An interesting addendum to the story of William Hart and the Fredonia Gas Light Company relates to the geological formation in which Hart originally found natural gas. That formation is known as the Marcellus Shale Play and extends thousands of square miles underground from mid-New York State into mid-Tennessee and from mid-Ohio to the Pennsylvania–Maryland border. Geologists believe that the formation may hold trillions of cubic feet of natural gas, enough to supply a significant fraction of the United States' energy needs for decades into the future. The most likely method for recovering that resource, however, is hydraulic fracturing, or "hydrofracking," a technology about which environmentalists and residents of the area have significant concerns. Whether the risks posed by hydrofracking are with the benefits provided by the huge supplies of natural gas is a question still being debated in the early twenty-first century.

Independent Petroleum Association of America

1201 15th Street NW, Suite 300
Washington, DC 20005
Phone: (202) 857-4722

Fax: (202) 857-4799
URL: http://www.ipaa.org/
Email: no address provided

The Independent Petroleum Association of America (IPAA) was founded in 1929 at a time of considerable uncertainty about domestic oil production in the United States. Some experts in the field of petroleum suspected that the United States had only a relatively limited supply of crude oil, and they were pushing the federal government to conserve the resources that were available and to promote in importation of petroleum from other parts of the world. American petroleum companies began to worry that these concerns would lead to the federal government's imposing limitations on the amount of crude oil that could be drilled in the United States, producing a devastating effect on their own businesses. It was within this climate, then, that a leader of the U.S. petroleum industry, Wirt Franklin, spoke out at a federally sponsored Oil Conservation Conference in Colorado Springs in June 1929. He warned that the drift toward importation of oil and conservation at home "would mean the annihilation and destruction of the small producer of crude oil" if it were allowed to come to pass. He suggested creating a new organization to protect and promote the development of domestic oil resources in the United States. That organization came to be called the Independent Petroleum Association of America.

It formally came into being on June 11, 1929, at a meeting that included about two dozen leaders of the U.S. petroleum industry, including not only a number of independent specialists in the field, but also representatives of major firms such as the Marine Oil Company, the Jennings-Heywood Oil Syndicate, the Simpson-Fell Oil Company, the Okmulgee Producers and Manufacturing Gas Company, and the Stanolind Oil and Gas Company, as well as media representatives of the Tulsa World and National Petroleum News publications.

Today, the IPAA claims to be responsible for the development of 95 percent of all domestic oil and gas wells and for the production of 54 percent of all domestic oil and 85 percent of all domestic gas. From the name of the organization, one might assume that IPAA members are primarily smaller companies who operate on the fringes of the primary major oil-producing system in the United States. It is not possible to confirm that impression, however, as IPAA apparently does not release its membership list to the general public. According to one outside study of the organization, 86 of the 100 top oil producers in the United States are officially designated as "independent" companies. Fifty six of those companies have market capitalizations of more than $1 billion, and the median market capitalization for the 86 independent companies is more than $2 billion. The top five independent oil companies in the United States are Apache, Anadarko, Devon, EOG and Talisman, with a total market value of more than $178 billion (Oil Change International; http://priceofoil.org/content/uploads/2012/03/FIN.OCI-Fact-Sheet-US-Subsidy-Removal-2012.pdf. Accessed on July 15, 2014).

In addition to its members, IPAA maintains close working relationships with about four dozen "cooperating associations" that includes a number of state and regional oil and gas associations, regional royalty owners associations, energy groups, and equipment suppliers. According to its mission statement, IPAA is "dedicated to ensuring a strong, viable domestic oil and natural gas industry, recognizing that an adequate and secure supply of energy is essential to the national economy."

One of the most important functions of the IPAA is one that carries over from its founding 85 years ago, namely working for the interests of the oil and gas industry in the United States, attempting to shape federal policy in such a way as to increase the exploration for and extraction of domestic oil and gas products. The organization carries out a number of activities in pursuit of this objective, including testimony before committees of the U.S.

Congress and a variety of federal agencies, providing comments to regulatory agencies on proposed rule-making and changes, and the development and distribution of fact sheets for policymakers and the general public on important national energy issues. Copies of some of these materials are available on the IPAA website.

A second important function of IPAA involves the collection, analysis, and distribution of data about oil and gas production in the United States. Some of its most important reports are the annual "Oil & Gas Producing Industry in Your States," which provides detailed statistics about the petroleum industry in each state; an important 2011 report on the contribution of independent onshore oil and gas producers to the U.S. economy; occasional profiles of independent oil and gas producers in the United States (the most recent of which dates to 2012–2013); an extremely useful collection of statistics on virtually every aspect of oil and gas production in the United States; and links to a number of international organizations and sources of information about worldwide statistics and the status of oil and gas production around the world.

IPAA also cooperates with the Petroleum Equipment Suppliers Association (PESA) in the support and operation of the Energy Education Center (EEC), founded in 2006 to introduce students to the fields of study needed to go forward with a career in the field of oil and gas recovery. The EEC currently operates geoscience, engineering, and leadership programs at five high schools in Houston and Fort Worth. The program's goal is to develop young people with the skills needed to move into a growing industry with an aging workforce.

A major news outlet for IPAA is its weekly newsletter "Issues and Insight," which focuses on important regulatory and related issues, Congressional and other federal and state legislative activities, and IPAA events. The organization also provides daily updates of important IPAA mentions nationwide and press releases on items and issues of interest to members of the organization and other observers.

Anthony R. Ingraffea (1947–)

The issue of hydraulic fracturing is frequently characterized today by vigorous debates between proponents and opponents of the practice, debates that often become acrimonious and even offensive. Yet, the fact is that well qualified experts in the field of petroleum engineering themselves often have very different views as to the safety, efficiency, necessity, and desirability of using fracking to retrieve fossil fuel resources from the Earth. Discussions and debates between experts on both sides of the issue are common and can be useful if participants are able to reserve their comments to the facts at hand. One of the most common academic spokespersons to speak in opposition to fracking is Cornell University professor Anthony R. Ingraffea. Ingraffea has written and spoken widely about the hazards that accompany the use of fracking to recover oil and natural gas supplies, but he is perhaps best known as lead author of an important 2011 study on the greenhouse gas footprint of methane gas, the primary component of natural gas. Ingraffea and his colleagues found that methane is a more dangerous greenhouse gas than either conventional gas or oil over any measured time period, and as much as 20 percent more harmful than coal in the short term and up to twice as harmful as coal over a 20 year period. These results are of considerable interest because proponents of natural gas often use its purported lesser effect on global climate change than other fuels.

Ingraffea has published well over 200 articles in refereed and specialized publications on hydraulic fracturing and a variety of other topics and has been the author or coauthor of more than 60 research reports on similar topics. He has been principal investigator or participant in more than 80 funded research projects in the fields of structural engineering, geotechnical engineering, engineering education, and cooperative research. Ingraffea is in great demand as a speaker on topics related to fracking, often as a participant in discussions and debates on the risks and benefits of the technology.

Anthony Richard Ingraffea was born on April 4, 1947, in Easton, Pennsylvania. After graduating from high school, Ingraffea enrolled at Notre Dame University with plans to become an astronaut. As the first step in that journey, he was awarded his BS in aerospace engineering from Notre Dame in 1969. He then took a job at Grumman Aerospace Corporation, still with plans to begin his astronaut training before long. After two years at Grumman, however, "things happened," as he later told interviewer Ellen Cantarow in a piece for EcoWatch. Among those "things" were the Vietnam War and the growing energy crisis in the United States. For a variety of reasons, then, he shifted his priorities and served, instead of in space travel, in the Peace Corps. From 1971 to 1973 he worked in Bejuma, Venezuela, as a county engineer responsible for all technical services in the jurisdiction.

Upon his return to the United States, Ingraffea enrolled at the University of Colorado at Boulder to pursue his doctorate in civil engineering, a degree he earned in 1976. Ingraffea's field of interest at Colorado was crack propagation in rock, a topic that was to become basic to the later development of fracking technology by the petroleum industry. After spending an additional year at Colorado as an instructor, Ingraffea accepted an appointment as assistant professor in the department of structural engineering at Cornell University, where he has remained to the present day. He worked his way up the academic ladder as associate and full professor, finally being named Dwight C. Baum Professor of Engineering in 1993, a post he continues to hold. During the academic year 1983–1984, Ingraffea took a sabbatical leave to work at the Lawrence Livermore National Laboratory, where he studied rock fracture stimulation. Since 2004, Ingraffea has also been visiting scientist at the Wright Patterson Air Force Base working in the field of structural sciences.

Ingraffea has said that he became interested in fracking as a political and social issue in about 2009 when he came across a number of newspaper articles, television advertisements, and other public pronouncements by the petroleum industry which, he told Cantarow, were "astoundingly inaccurate." He said that he felt some responsibility for setting the record straight on the potential risks of hydraulic fracturing and found that he was soon in wide demand to speak about the technology. In recognition of his contributions both in the field of academia and as a spokesperson about fracking issues, Ingraffea has been awarded a number of honors, including the Cornell School of Civil Engineering Professor of the Year award for 1977–1978, National Research Council/U.S. National Committee for Rock Mechanics 1978 Award for Outstanding Research in Rock Mechanics at the Doctoral Level, Cornell College of Engineering Professor of the Year award for 1978–79 and for 1981–1982, Dean's Prize for Innovation in Teaching, Cornell College of Engineering, National Research Council/U.S. National Committee for Rock Mechanics 1991 Award for Applied Research, Honor Award, University of Notre Dame, College of Engineering, for Significant Contributions to the Advancement of Engineering, George R. Irwin Medal, American Society for Testing and Materials, and Fellow, International Congress on Fracture.

International Energy Agency

9, rue de la Fédération
75739 Paris Cedex 15
France
Phone: (33-1) 40 57 65 54
Fax: (33-1) 40 57 65 59
URL: http://www.iea.org/
Email: info@iea.org

The International Energy Agency (IEA) was founded in 1974 in response to the world crisis in energy supplies resulting from the oil embargo announced by the Organization of Arab Petroleum Exporting Countries (OAPEC). The embargo had been established as a way by which Arab states could object to support of the state of Israel by the United States and other Western nations. Confronted with the sudden and drastic loss of its most important energy source, petroleum, a number of nations banded together to form the International Energy Agency. IEA's initial goals were to develop more sophisticated and accurate statistics on known and estimate petroleum reserves and oil production and consumption and to investigate ways of ameliorating the effects of the dramatic loss in petroleum supplies for developing nations. Since 1974, IEA's mission has expanded somewhat and now includes four major themes.

The first of these themes is Energy Security, which involves the promotion of diversity and flexibility of energy supplies in order to avoid the worst consequences of fuel crises like that of the 1974 oil embargo. The second theme is Environmental Protection, which recognizes the deleterious effects of extensive fossil fuel use. A recent feature of this theme is a greater emphasis on the possible global climate effects resulting from the combustion of coal, oil, and natural gas. The third theme is Economic Growth, which emphasizes the importance of developing energy policies that ensure the continued economic development of all nations, whether developed or developing. The fourth theme is Engagement Worldwide, which is a specific acknowledgment that the energy and environmental policies facing nations of the world transcend national borders and can only be solved by international cooperation.

Only member states of the Organisation for Economic Co-operation and Development (OECD) are permitted to belong to the IEA. Current members are Australia, Austria, Belgium, Canada, Czech Republic, Denmark, Finland, France, Germany, Greece, Hungary, Ireland, Italy, Japan, Luxembourg, Netherlands, New Zealand, Norway, Poland, Portugal, Slovakia, South Korea, Spain, Sweden, Switzerland, Turkey, the United Kingdom, and the United States. Although prohibited from inviting non-OECD nations to join the agency, the IEA has a well-developed program for working with such nations through its Directorate of Global Energy Dialogue (GED). Established in 1993, the GED attempts to better understand the status of energy systems in non-OECD nations and to share with those nations some of the knowledge and expertise developed through IEA programs. The agency currently has bilateral and regional agreements with a number of non-OECD nations and regions to achieve these goals, including Brazil, Caspian and Central Asia, Central and Eastern Europe, China, India, Mexico, the Middle East, Russia, Southeast Asia, Ukraine, and Venezuela.

The primary decision-making body of the IEA is the Governing Board, which consists of the energy ministers of all member states, or their designated representatives. Decisions made by the Governing Board are carried out by the Secretariat, which consists of researchers and other scholars. The Secretariat's work involves the collection of data on energy supplies, production, and consumption; sponsorship of conferences and other meetings on energy issues; assessment of energy conditions in member states; development of projections for global energy futures; and proposals for national energy policies in member states.

The work of the Secretariat is organized around a number of specific topics, for which a variety of reports and publications are generally available. The current topics of interest to the IEA include coal, carbon dioxide capture and storage, cleaner fossil fuels, climate change, electricity, energy efficiency, energy indicators, energy policy, energy predictions, energy statistics, fusion power, greenhouse gases, natural gas, oil, renewable energy, sustainable development, and technology. For the general public and experts in the field, IEA's most important contributions may well be their regular reports on a variety of important energy-related issues. In the area of statistics, the agency publishes a host of reports on a monthly, quarterly, and annual basis. Perhaps best known of these is its annual *Key World Energy Statistics*, published regularly for the last

10 years. Other statistical publications deal with energy prices and taxes, energy statistics for OECD and non-OECD states, oil information, natural gas information, renewable information, and carbon dioxide emissions from fuel combustion. The agency also publishes a quarterly publication, *Oil, Gas, Coal, and Electricity*.

Other fields for which publications are available are Global Energy Dialogue, which includes publications on energy issues in non-OECD nations; Energy and Environment, which deals with topics such as the environmental and climate effects of fossil fuel combustion; Renewable Energy, which considers technical, social, economic, and political aspects of the development of renewable energy sources; Energy Efficiency, which discusses ways in which conservation can extend the lifetime of existing fossil fuel resources; Energy Technology Network, which reports on agreements made among member states to improve their energy production and consumption patterns; Policy Analysis and Cooperation, which reviews energy policies, practices, and projections for specific member states; and Energy Technology, which reviews proposed and in-place changes in technology to make more efficient use of existing and proposed energy sources. An especially interesting feature of the agency's website is its "Fast Facts" section, which lists tidbits of information about many aspects of energy. Each item is cited from an IEA publication, to which a link is provided in each case.

In addition to its print publications, the IEA sponsors a number of conferences and other meetings, provides speakers for a variety of professional and community events, and testifies before governmental agencies in member nations. In 2014, for example, the agency sponsored, cosponsored, or participated in meetings on bioenergy opportunities in southern Africa, in Durban, South Africa; the future of nuclear technology, in Paris; a forum on unconventional gas, in Calgary, Canada; international standards on renewable energy and energy efficiency, in Paris; hydrogen technology futures, in Bethesda,

Maryland; and policies for mitigating climate change, in Warsaw.

Interstate Natural Gas Association of America

20 F Street, NW, Suite 450
Washington, DC 20001
Phone: (202) 216-5900
URL: http://www.ingaa.org/
Email: http://www.ingaa.org/Contact.aspx

The Interstate Natural Gas Association of America (INGAA) was founded in 1944 as the professional association of companies that build and operate pipelines that carry natural gas from the source of production to consumers, exporting sites, and other locations at which the natural gas is put to use. INGAA currently consists of 25 members who are responsible for the vast majority of natural gas pipelines in the United States and Canada. Those pipelines extend a combined distance of more than 200,000 miles in the two nations. Some of the INGAA members are Alliance Pipeline, Carolina Gas Transmission Corporation, Iroquois Pipeline Operating Company, Pacific Gas and Electric, Piedmont Natural Gas, and Sempra US Gas & Power.

The INGAA Foundation is a sister organization created in 1990 to promote the use of natural gas by the consuming public in an environmentally safe manner. The foundation has more than 120 members that include a variety of pipeline-related businesses, including gas pipeline companies, construction companies, engineering firms, pipe and compressor manufacturers, accounting firms, companies providing information technology services, and other suppliers of goods and services to the pipeline industry. The foundation funds studies and reports and sponsors forums on technical and economic issues that are of interest to its members.

An important part of INGAA's work is educating the general public and others about pipelines and their impact on the natural environment and human health. The Pipelines 101 section of the association's website provides such an introduction, with sections on basics of natural gas, landowner leases and commitments, pipeline construction, pipeline operations, pipeline regulation, pipeline safety, and pipeline "fun facts."

The issues surrounding pipeline construction and operation have become entangled in the debate over hydraulic fracturing in recent years because of plans to ship a significant portion of the natural gas produced by fracking to foreign nations, especially those in Asia. That process involves moving natural gas hundreds of miles from the point of production to treatment facilities, usually located on the eastern, western, or Gulf of Mexico coasts of the United States. At the treatment facilities, the gaseous product is converted to a liquid, LNG, which can then be transported by ship to consumers overseas.

The growing demand for LNG for export in the past decade has meant that companies are having to build hundreds of miles of new pipeline. In many cases, that pipeline has to run through private property that can be leased or purchased for that purpose only by negotiations with a landowner or by means of eminent domain. Such negotiations have often resulted in acrimonious disputes between pipeline companies and landowners involved in such transactions, adding one more layer of controversy over the expanding use of fracking for the recovery of natural gas.

One of the primary objectives of INGAA is providing information about safety issues relating to the use of natural gas pipelines. The organization's website contains a useful section that deals with specific issues such as first responses to pipeline accidents by pipeline companies, instructions about what to do in case of an emergency, pipeline markers (markers that identify the characteristics of any given pipeline), regulation of pipelines, and specific safety activities provided by the pipeline industry and individual companies. INGAA also offers safety workshops for member organizations on specific safety practices related to pipeline operation.

Both INGAA and the INGAA Foundation also produce a large number of technical and general reports on a host of topics relating to pipeline construction and operation. All of these reports are available for download at no charge at the INGAA website, http://www.ingaa.org/11885/Reports.aspx. Some of the topics included in the report series are The Role of Pipeline Age in Pipeline Safety, North America Midstream Infrastructure through 2035: Capitalizing on Our Energy Abundance, Planning Guidelines for Pipeline Construction during Frozen Conditions, Building Interstate Natural Gas Transmission Pipelines: A Primer, and Validation of Direct Natural Gas Use to Reduce CO2 Emissions.

INGAA also publishes a series of position papers entitled INGAA on the Issues dealing with topics of special interest and, usually, controversy. The association's paper on LNG exports is of special interest to those concerned with the movement of fracked natural gas across the continent (http://www.ingaa.org/issues/18035.aspx).

Interstate Oil and Gas Compact Commission

900 NE 23rd St.
Oklahoma City, OK 73105
P.O. Box 53127
Oklahoma City, OK 73152
Phone: (405) 525-3556
Fax: (405) 525-3592
URL: http://iogcc.publishpath.com/
Email: iogcc@iogcc.state.ok.us

The Interstate Oil and Gas Compact Commission (IOGCC) was created in 1935 by the governors of six states who were concerned about the lack of regulation over petroleum production and the consequent surplus of product this condition had engendered. Earlier efforts by individual states to work together on this issue were discouraged largely by the petroleum industry itself, which preferred the somewhat laissez-faire approach of the federal government to its operations. Finally in 1935, however, E. W. Marland, then governor of Oklahoma and former U.S. representative from the state, was able to use federal authority to allow individual states to band together to form a *compact* agreement through which they could work together on common problems. The organization's original charter was signed on February 16, 1935, in Dallas, by representatives of the six states, Colorado, Illinois, Kansas, New Mexico, Oklahoma, and Texas. The IOGCC currently consists of 30 member states; 8 associate member states; and 12 international affiliates in Canada, Egypt, Georgia, and Venezuela. The organization also has associate relationships with the U.S. Department of the Interior, Department of Energy, Environmental Protection Agency, Federal Energy Regulatory Commission, and National Association of Regulatory Utility Commissioners.

The primary function of the IOGCC is to assist states in developing regulatory policies and practices that ensure the sound development of oil and gas resources while also ensuring the health and safety of citizens as well as the quality of the national environment. The organization works through a variety of committees and working groups that consist of state governors, state oil and gas regulators, and representatives of industrial and environmental groups to identify and make use of best practices in the field to achieve these objectives. Identifying and disseminating information on best practices is accomplished by surveying practices currently in use in specific states; cataloging and publishing this information in the form of reports, briefs, and other types of printed and electronic media; and exchanging this information with states in a variety of ways, such as biannual meetings of stakeholders interested in the regulation of oil and gas operations. An important feature of the organization's work is an annual conference that provides an opportunity for the sharing and dissemination of information on new and promising practices and policies. The 2014 and 2015 conferences were planned for Columbus, Ohio, and Oklahoma City, respectively. IOGCC also offers a number of midyear meetings of various types, such as the 2011 Marcellus Summit in State College, Pennsylvania; 2012 Midyear Meeting in Vancouver, British Columbia; and 2013 Midyear Issues Summit in Point Clear, Alabama.

Eight specific issues currently dominate the activities of the IOGCC: states' rights, national energy policy, hydraulic fracturing, environmental stewardship, model statutes, carbon sequestration, produced water, and manpower shortages. The IOGCC website provides detail about the background of each of these issues and describes the activities being undertaken by the organization in an effort to deal with the issues. In the area of hydraulic fracturing, for example, the IOGCC website provides a brief background on the technology, information on state and congressional actions, a summary of IOGCC statements and resolutions, and a link to FracFocus, a website cosponsored with the Ground Water Protection Council on the chemicals used in the fracking process.

IOGCC currently has about a dozen publications available in both print and electronic format on a number of issues related to the regulation of oil and gas operations. These reports deal with topics such as marginal wells, legal and regulatory issues related to carbon dioxide storage, reducing the environmental footprint of oil and gas extraction, adverse impact reduction, Rocky Mountain crude oil market dynamics, and current manpower status and issues in the petroleum industry. The organization also published other products such as press releases, a media kit on relevant issues, and a regular newsletter available at no charge electronically on the IOCGG website (http://iogcc.publishpath.com/newsletter).

E. W. Marland (1874–1941)

Marland was the primary force behind the creation in 1935 of the Interstate Oil and Gas Compact Commission (IOGCC), established to develop a method of controlling the production of crude oil in the United States. The early 1930s were a difficult time for the petroleum industry. An approaching economic depression resulted in a vastly decreased demand for crude oil with the result that the price of a barrel of oil dropped to its lowest point in history, in the range of 10 cents per barrel. At the time the federal government left oil production policy largely in the hands of petroleum companies, each of whom had its own expectations and plans for production. Concerned by the lack of any central authority for the management of oil production, distribution, and sales in the United States, Marland called a meeting of governors from six states (including his own of Oklahoma) at his home in Ponca City, Oklahoma, to set up a commission to deal with this problem. The solution to the problem faced by the governors was provided by congressional authority that allowed states to band together in *compacts* through which they could work jointly to solve common problems. At the conclusion of the Ponca City meeting, the governors agreed to form the Interstate Oil and Gas Compact Commission (IOGCC) with the mission of creating an administrative mechanism for the control of crude oil production in the United States. That commission remains in existence today, continuing in its long history of monitoring petroleum production in the United States.

Ernest Whitworth Marland was born in Pittsburgh, Pennsylvania, on May 8, 1874, the youngest child of eight and the only boy born to Alfred and Sara Marland. He attended Park Institute in Pittsburgh and then enrolled at the University of Michigan, where he studied the law. He graduated with his LLB degree from Michigan in 1893 at the age of 19 and returned to Pittsburgh to begin his practice of the law. The late nineteenth century was a time of change and challenge in Pennsylvania with the nation's first commercial oil well having been drilled in Titusville only two decades earlier. As with many new industries, great opportunities were available for adventurers who were willing to take their chances on a new technology. Marland was such a person, and he had made his fortune in oil speculation by the time he was 30 years old. As is also the case with many technologies, however, Marland also lost his fortune nearly as quickly as he earned it, largely as the result of the great financial panic of 1907.

Undiscouraged by his misfortunes, Marland simply borrowed money and moved to Oklahoma, where the promise of new oil finds had just been announced. He settled in Ponca City and soon was on his way to a second fortune. After forming a series of companies to explore for oil, he consolidated them all in 1921 under the name of the Marland Oil Company. His success was short-lived, however, as he gradually lost control of his own company to the banks from which he had borrowed so much money. By the early 1930s, Marland Oil no longer existed and had been replaced by a new corporation, the Continental Oil Company, forerunner of today's ConocoPhillips, one of the largest oil producers and refiners in the world.

As he was gradually displaced from his role in Marland/Conoco, Marland became more interested in politics and decided to run for the U.S. House of Representatives from Oklahoma's Eight Congressional District (which no longer exists). As a liberal Democrat, Marland supported Franklin D. Roosevelt's progressive program for the nation and was swept into office with him in the 1932 elections. Marland served only one term in the House of Representatives, however, as he chose to run for governor of Oklahoma instead in 1934. He won that election and served only a single term, after which he attempted to revive the Marland Oil Company, an effort in which he was unsuccessful. In 1940, he chose to run again for the U.S. House of Representatives, but he lost that election and returned to his home in Ponca City. He survived this latest loss by only a year, however, as he died on October 3, 1941.

In 2012, the Weinstein Company announced plans to make a film about the somewhat notorious relationship between Marland and his adopted daughter, Lydie, that some people have said led to the downfall of the Marland Oil empire. (As of late 2014, the film had not yet been made.)

Aubrey McClendon (1959–)

McClendon was cofounder, with Tom L. Ford, of Chesapeake Energy, currently the second largest producer of natural gas in the United States. McClendon and Ford founded Chesapeake in 1989 with $50,000 in cash, eight employees, and no oil or gas leases during an economy in which oil and gas prices were hovering around their lowest prices. The company focused on the newly developed technology of horizontal drilling in unconventional sites and soon experienced a successful run of strikes that brought economic security to the company. Between 1994 and 1996, Chesapeake reported the highest growth rate of any company among 8,000 oil and gas corporations in the United States. Efforts to extend their practice into parts of Texas and Louisiana in the late 1990s proved to be unsuccessful, however, and the company returned to its roots in Oklahoma, where it decided to focus on exploration for and extraction of natural gas, primarily with its horizontal drilling technology. The company also vastly extended its land holdings with the aim of making it the nation's largest producer of natural gas, a goal they have almost achieved. In 2008, McClendon was the highest paid CEO in the S&P top 500 companies with a total income of $112 million. He has also been named to *Forbes* magazine's CEO 20–20 Club for having produced at least 20 percent in profits over 20 years or more of service.

Aubrey Kerr McClendon was born on July 14, 1959, in Oklahoma City. His middle name provides a clue to the illustrious family into which he was born. He was a grand-nephew of Robert S. Kerr, former governor of Oklahoma and United States senator from the state from 1949 to 1963. Kerr was also cofounder of Kerr-McGee Oil Industries, one of the largest petroleum companies in the world during its heyday. McClendon's father, Joe, worked for Kerr-McGee for 35 years in what biographer Gregory Zuckerman has called "the dull and dreary side of the business," at a desk job in the refining and marketing division of the company. With that background, young Aubrey grew up with no particular passion for the petroleum industry and decided to major in history when he matriculated at Duke University in 1977.

Ignoring his college major, McClendon accepted a job at the accounting firm of Arthur Andersen in Oklahoma City after graduation. He never actually started work at the company, however, as he had a better offer from an uncle, Aubrey Kerr, Jr., for a similar job at better wages in a petroleum company, Jaytex Oil & Gas, owned by Kerr. McClendon worked in the Jaytex accounting department for only nine months, however, before he decided that he wanted more exciting work than that provided in the accounting department. In actuality, he was looking not only for more interesting work but also for an opportunity to make his fortune. As a step in that direction, he moved to the firm's land department where he took a job as a landman. A landman is a person in the petroleum industry who has a wide range of responsibilities relating to the identification of lands that might be productive for drilling operations, finding out who owns those lands, and negotiating for mineral rights, rental, or purchase of the land for the company's use. A talented landman can be of enormous value to a petroleum company because he or she (and it's usually "he") is the key to finding locations where the company can drill and presumably start making a great deal of money on land that it owns, rents, or at the very least, on which it pays royalties for mineral rights.

While he was traveling around the region in his work as a landman, McClendon came to know Tom Ford, who was also just beginning a career in the petroleum industry. The better the two men got to know each other, the more they realized that they could be a successful team, and in 1983 they decided to join efforts and form their own company. They decided to name the company after McClendon's favorite vacation spot, Chesapeake Bay. McClendon and Ford proved to be an efficient team with skills that nicely complemented each other. Over the next three decades, Chesapeake experienced its ups and downs, but ultimately became the premiere natural gas producing company in the United States.

In 2012, shareholders of Chesapeake Energy launched what one business writer, David Benoit, has called a "historic revolt" against McClendon's leadership. The major complaint seemed to be that McClendon was still running the company as his own personal fiefdom, as he had done before the company went public in 1993. McClendon was allowed to keep his title as chief executive officer, while relinquishing the title of chairman. He gave up even that title a year later when he left Chesapeake completely, forming his own new company, American Energy Partners.

Throughout most of his career, McClendon has been a forceful and articulate spokesperson for natural gas. He has touted the economic and environmental advantages of the fuel, especially in comparison to those of coal and oil. As part of this effort, he was one of the founding members of America's Natural Gas Alliance in 2009. In addition to his work in the petroleum industry, McClendon is an avid sports fan and is currently an owner of the Oklahoma City Thunder professional basketball team.

George Mitchell (1919–2013)

Mitchell has been described as "the father of fracking." That term is almost certainly overgenerous as he did not invent the process, and fracking had been widely used in petroleum exploration for years before he appeared on the scene. Mitchell is given such high credit, instead, for his other contributions to the field of hydraulic fracturing. One of those contributions was his perseverance in searching for natural gas in regions where other companies had long before given up or had decided the search was not even worthwhile. For example, Mitchell's greatest breakthrough may well have been his decision to drill into a layer of shale in the vast Barnett Play in the 1990s even though a number of other companies had decided such a venture was doomed to failure. Mitchell also took the revolutionary notion of using a very different type of drilling fluid that other companies had used in the past, substituting a mixture of primarily water and sand (with some added chemicals) in place of the traditional water mixed with gels and foams that had been used almost universally. In any case, his decision to keep trying to find natural gas in places where few others had ever looked (or had given up looking) with technologies that had seldom before been tried were eventually hugely successful. He is credited with having developed more than 10,000 successful wells during his lifetime, bringing enormous success to his company, Mitchell Energy, which he eventually sold to Devon Energy in 2002 for $3.5 billion.

George Phydias Mitchell was born on May 21, 1919, in Galveston, Texas, to Savvas Paraskevopoulos, a native of the village of Nestani, Greece, and Katina Eleftheriou, also from a rural Greek town, Argos. Mitchell's father had been a sheepherder in Nestani before deciding to pursue a more promising future by immigrating to the United States at the age of 20. His story of walking across Greece to board a ship that was to take him to New York, where he set out to make his fortune hardly knowing the English language, is an inspiring story of perseverance and success that has been retold a number of times. (See, for example, *How Mitchell Energy and Development Corp. Got Its Start and How It Grew*, Universal Publishers, 2001, 363–366.) At one of his first jobs working on the railroad, the paymaster found Paraskevopoulos's name too difficult to spell, so assigned him a new name, the same as the paymaster's, of Mitchell.

George Mitchell grew up in Galveston, attending elementary and secondary schools there. He was a bright student and skipped a couple of classes in elementary school, resulting in his graduating from Galveston's Ball High School at the age of 16. He originally planned to attend Rice University to major in medicine (his mother's dream as much as his own), but he changed his mind during a one-year post-graduate program he pursued after leaving Ball. He especially enjoyed a math class he took, and a discussion about his college plans with his brother Johnny helped to change his mind. In the fall, he enrolled at Texas A&M, rather than Rice, and began a major in geology. He graduated four years later with a degree in petroleum engineering and a concentration in geology, a program that normally required five years to complete.

After receiving his degree, Mitchell worked for a short time for the Amoco company in Louisiana on a drilling rig before enlisting in the U.S. Army when World War II broke out. During the war he served in the Corps of Engineers at its base in Houston throughout the war. After the war ended, Mitchell founded his own business as an oil and gas consultant in Houston, where one of his first customers was a small drilling company called Roxoil Drilling. Within a relatively short period of time, he bought the company for $9,000 and changed its name to Oil Drilling. Oil Drilling was the business venture through which Mitchell was to carry out his many and varied activities over the next half century. In 1971, the company was renamed the Mitchell Energy and Development Corporation. Mitchell's success in the oil and gas business was largely a result of his unwillingness to give up on ideas in which he had confidence, even if they had been unsuccessful for others. In looking back on his career, the newspaper *Economist* observed in 2012 that Mitchell

> saw the potential for improving a known technology, fracking, to get at the gas. Big oil and gas companies were interested in shale gas but could not make the

> breakthrough in fracking to get the gas to flow. [He succeeded in spite of the fact that] Everyone . . . told him he was just wasting his time and money. ("Gas Works." The Economist. July 14, 2002. http://www.economist.com/node/21558459. Accessed on June 10, 2014.)

Mitchell is probably as well known today as an environmentalist and philanthropist as an oil and gas magnate. In the mid-1970s, Mitchell became concerned about the way in which human development was despoiling the natural environment and decided to explore the possibility of building a new community that would exist in consonance with the environment. In 1974, he initiated just a community, The Woodlands, built on 44 square miles of land 28 miles north of Houston. Mitchell envisioned an environmentally friendly mix of homes, schools, retail malls, office parks, distribution centers, golf courses and other recreational facilities, and a large conference center for a mixed population ranging from wealthy to low-income residents. Although disparaged by some business-savvy observers at its birth, the community continues to thrive today with an estimated 2013 population of 107,769. Mitchell and his wife, Cynthia, also contributed to a wide variety of educational environmental, and science-related institutions and activities, having spent an estimated $400 million to date on such activities.

Mitchell died at his home in Galveston on July 26, 2013, at the age of 94.

The National Association of Royalty Owners

15 W. 6th St., Suite 2626
Tulsa, OK 74119
Phone: (918) 794-1660
Toll-free: (800) 558-0557
Fax: (918) 794-1662
URL: http://www.naro-us.org/
Email: NARO@NARO-us.org

The National Association of Royalty Owners (NARO) was founded in Oklahoma City in June 1980 by a group of about 500 landowners from 10 states. The group had come together in response to the recent adoption by the U.S. Congress of a windfall profits tax designed originally to increase the tax on large petroleum companies that were making enormous profits at the time. A somewhat unexpected consequence of the act was that landowners who had leased a portion (or all) of their property to such companies were also ensnared in the provisions of the law, and many saw their royalties from leases drop dramatically, in some cases by as much as 70 percent. Since these landowners had originally negotiated their leases in good faith with petroleum corporations, they had expected to be able to live off the income from royalties in the future. When they learned that such was likely not to be the case as a result of the windfall profits tax, they revolted, ending up with the formation of NARO as a mechanism through which they could more effectively express their concerns about the issue. The leader of the landowner rebellion was James L. Stafford, executive vice president of the public relations firm ICPR at the time the tax act was passed. Stafford was elected the first president of NARO, a position he held until 2000. Stafford had become involved in the landowner rebellion because he himself was a landowner and a recipient of oil and gas royalties.

Today, NARO calls itself "the ONLY national organization representing, solely and without compromise, oil and gas royalty owners interests." The organization attempts to represent the interest of an estimated 8.5 million royalty owners. It describes the average such individual as a 60-year-old widow who receives less than $500 a month from her lease agreement. NARO is run by a board of directors consisting of the president, vice president, corporate secretary, treasurer, immediate past president, five at-large members, and the presidents of the 10 "state" chapters (Appalachia, Arkansas, California, Louisiana, New York, North Dakota, Oklahoma, Pennsylvania, Rockies, and Texas). Each of the state/regional chapters is relatively independent, often with its own statement of mission, and with its own slate of officers, schedule of events, section on news from its own district, and specific set of issues with which it is concerned. The NARO-New York web page on the NARO website, for example, provides its mission statement; list of coming events; information on natural gas, royalties, and leasing; and its own videos and news section.

NARO offers various levels of membership, beginning with a basic membership at the cost of $150 per year. A business, corporate, and institutional membership is available at the rate of $550 per year, and a family trust membership valid for up to four members can be had for $300 per year. A Golden Eagle membership in the NARO Foundation Eagles is $1,500 per year. Information about the NARO Foundation, a 501(c)(3) charitable corporation, is not readily available.

One of the most useful parts of the NARO website is a collection of links to associations, agencies, and organizations of interest to the landowner or leaseholder, including entities such as the federal government, nearly all state governments, gas and oil companies and associations, and a number of media outlets. The organization's event calendar carries listings not only of its own national conventions, conferences, and other meetings but also those of the state and regional divisions of NARO.

No Fracked Gas in Mass

Berkshire Environmental Action Team
27 Highland Ave.
Pittsfield, MA
Phone: (413) 230-7321
URL: http://www.nofrackedgasinmass.org/
Email: nofrackedgasinmass@gmail.com

As noted in the introduction to this chapter, No Fracked Gas in Mass is typical of the dozens of state and local groups that have been organized to protest fracking or some related activity within their specific geographic range. Other groups that fall into this category are Cook Inletkeeper (Alaska), Arkansas Fracking.org, Whittier Hills Oil Watch (California), Erie Rising (Colorado), Idaho Residents against Gas Extraction, Frack Free Illinois, Don't Frack Michigan, Frack Free Nevada, Rochester Defense against Fracking (New York), SW Ohio No Frack Forum, Stop Fracking around Chattanooga (Tennessee), and Powder River Basin Resource Council (Wyoming).

No Fracked Gas in Mass is a project of the Berkshire Environmental Action Team (BEAT), which organized in 2002 when a small group of citizens in the Berkshire area realized that existing laws and regulations were being ignored in efforts to protect a body of water in their area. The organization has grown over time to extend its efforts to a wide variety of environmental issues affecting western Massachusetts and adjacent areas. No Fracked Gas in Mass is one of those current projects. The campaign was started in response to an announcement by the Tennessee Gas Pipeline Company (a subsidiary of the Kinder Morgan company) that it planned to build a pipeline running from Richmond, in Berkshire County, across Massachusetts to a shipping terminal to be built at Dracut, north of Boston. The campaign has also adopted a broader mission "to stop the expansion of fossil fuel infrastructure in Massachusetts and to promote expanded efficiency and sustainable, renewable sources of energy and local, permanent jobs in a clean energy economy."

The proposed pipeline is one of the ancillary projects being developed in connection with the rapidly growing effort to extract natural gas from shale formations in various parts of the nation. Companies plan to mine natural gas not only to meet domestic needs for clean fuels of the future, but also to export to nations around the world, especially those in Asia, with energy needs they cannot meet with their own resources. In the case of natural gas, the fuel is typically shipped from the extraction site, almost always at a considerable distance from the coast, to an ocean port, where it is then converted to LNG, a form in which it can be shipped by tankers. Companies have been aggressive in obtaining the rights of way they need to build such pipelines, often causing disruption of local communities and individual properties. Residents of affected area worry not only about the possibility of leaks and possible explosions from the pipelines, but also a general disturbance of their personal and community lives.

No Fracked Gas in Mass encourages members and other concerned citizens to take action against the proposed pipeline (and against the process of fracking itself) on a number of levels, including federal, state, and local. It provides information on existing and proposed laws and regulations dealing with natural gas in an effort to inform people as to the options they have in responding to the perceived threats the pipeline poses to the Massachusetts community. It provides similar information about state existing and pending laws and regulations, as well as corporations operating within the state with a stake in the pipeline and fracking in general. No Fracked Gas in Mass also works specifically at the local level, where community laws and regulations are sometimes the best first line of action against environmentally insensitive projects. The organization provides links to local officials in areas likely to be impacted by the pipeline and by fracking.

One of the most useful services provided by No Fracked Gas in Mass is a page on its website summarizing news about activities in the six counties in Massachusetts and four counties in New York most severely affected by the proposed pipeline. For each of the 10 counties, an individual page provides links to all of the cities and towns in which actions have occurred or are being planned in response to the pipeline.

A particularly useful page on the organization's website is entitled The Score. It summarizes the achievements thus far racked up by the organization in its efforts to prevent the pipeline from being built. It includes the number of signatures that have been obtained on a petition to the state Office of Energy and Environmental Affairs, the Massachusetts State House, the Massachusetts State Senate, and Governor Deval Patrick, as well as copy of the petition itself for those who wish to sign it; the number of petitions adopted by towns against the pipeline, along with copies of those petitions; statements from office holders and individual citizens about the issue; a list of state legislators who support the work of the organization; and statements and petitions of support from other environmental groups.

A section of the No Fracked Gas in Mass website also provides a collection of "Latest News and Thoughts," which appears to make available almost everything that has been said or written about the projected pipeline and the subject of fracking in western Massachusetts and eastern New York. The page is a superb introduction to the history of the dispute over this project and an overview of the arguments that can and have been made against both natural gas pipelines and the operation of fracking itself. In a broader sense, the No Fracked Gas in Mass website also provides links to a number of other projects and organizations, both national and regional, concerned with pipeline and fracking issues and so is a good guide to learning about other projects of similar design and focus.

Edward A. L. Roberts (1829–1881)

Hydraulic fracturing is generally regarded as a largely twenty-first-century development. But the basic concept underlying fracking has been known for at least 150 years, with some of the earliest experiments on this technology dating to the Civil War era. One of the earliest proponents of the concept was Colonel Edward A. L. Roberts, who fought for the Union side in that war (until he was cashiered in 1863). In fact, Roberts got his idea for hydraulic fracturing (although it was not called by that name at the time) during the Battle of Fredericksburg in 1862. He saw members of the Confederate Army fire a cannon into a narrow canal that stood in the way of the troops' advance. The action blew away materials causing an obstruction in the canal, allowing water to flow freely through the new opening.

Roberts put this observation to work three years later in dealing with a problem common to oil wells. Over time, solid and semisolid materials present in petroleum tend to clog the pores through which oil flows into a well. The most common of these materials is paraffin, a waxy material present in small amounts in all crude oil. The paraffin effect is sometimes referred to as the scourge of oil drilling because it can dramatically reduce the flow of oil even when a reservoir still contains large amounts of crude oil and natural gas.

It occurred to Roberts (as it had to some of his predecessors in the petroleum industry) that it might be possible to use a small charge of explosives to produce the force needed to open up the pores in oil-bearing rock and allow the product to begin flowing again. He tested his idea with a device he called a *torpedo* on a well called Ladies Well on Watson's Flats south of Titusville, Pennsylvania on January 21, 1865. Roberts torpedo consisted of an iron case holding 15 to 20 pounds of explosive that could be lowered into a well to the point at which clogging was thought to exist. The torpedo was surrounded in the well hole by water, which buffered and concentrated the force of the blast. The explosive was then detonated by means of a cap at the top of the torpedo. The explosive used in the torpedo was either black powder or nitroglycerin. According to reports at the time, Roberts's device increased the flow of oil and gas into the well by up to 1,200 percent within the first week following the blast. He charged a fee of $100 to $200 plus 1/15th of the value of the increased flow of oil.

Edward A. L. Roberts was born in Moreau, Saratoga County, New York, on April 13, 1829. He enlisted in the U.S. Army in 1846 at the age of 17 and served in the Mexican War with distinction. After the war, according to one biographical sketch, he returned to Moreau and "entered an academy and passed several years acquiring a higher education." He then took a position in the dental office of his brother in Poughkeepsie, New York. The brothers later moved to New York City where they opened a business for the manufacture of dental equipment. One of the fruits of this collaboration was a series of patents for a variety of inventions in the field of dentistry, for which Edward Roberts was awarded a number of gold medals by the American Institute of the City of New York.

With the outbreak of the Civil War, Roberts returned to active duty and was commissioned as lieutenant colonel of the 28th New Jersey Volunteer Infantry on September 15, 1862. Shortly thereafter he made something of a name for himself at the Battle of Fredericksburg, where he first observed the use of explosives for clearing a waterway. A year later, he offered his resignation from the army, but instead was cashiered for "resigning in the face of the enemy." Returning to New York, Roberts and his brother began to think about the potential of the growing field of petroleum exploration and development and decided to move to the focus of that action, near Titusville, Pennsylvania. It was during this period that Roberts began to think about the use of explosives for the clearing of well holes and conducted his first experiment with the device that was to become known as the Roberts Torpedo. He received a patent for his invention in 1866 and formed the Roberts Petroleum Torpedo Company which, according to reports of the time, experienced "enormous success." In fact, its success was so great that the company soon became embroiled in a series of endless law suits attempting to protect their patent and business activities. Many of the disputes with which the company had to deal was the efforts of freelancing men who used their own versions of the torpedo to clear wells in the middle of the night—at much cheaper fees than Roberts charged—to avoid detection. The practice is said to have led to the now popular expression called *moonlighting*, during which individuals carry out illegal activities by the light of the moon. The Roberts sold their company in 1903 to Adam Cupler, Jr., who renamed it the Cupler Torpedo

Company. In 1937, the Cupler Torpedo Company was purchased by the Otto Torpedo Company, forming a company that remains in business today, the Otto Cupler Torpedo Company.

In any case, Roberts's invention led to both fame and fortune for him and his family. He is said to have built the Brunswick Hotel in Titusville, where his family resided for many years. He died on March 25, 1881 in Titusville.

The Endocrine Disruption Exchange

P.O. Box 1407
Paonia, CO 81428
Phone: (970) 527-4082
Fax: (970) 527-4082
URL: http://endocrinedisruption.org/
Email: http://endocrinedisruption.org/contactus

The Endocrine Disruption Exchange (TEDX) was founded in 2003 by Theo Colborn, who was then professor of zoology at the University of Florida at Gainesville. Colborn continues to serve as president of TEDX and as emerita professor at the University of Florida. TEDX grew out of Colborn's longstanding interest in the effects of chemicals—primarily synthetic substances—on the human endocrine system. Such substances can interrupt the normal function of the endocrine system at very long concentrations, sometimes at levels of less than one part per trillion. Although the topic of endocrine disruptors has been studied by a number of researchers, TEDX is reputedly the only organization in the world that focuses exclusively on the topic. (An important introductory document on the concept of endocrine disruptors is Colborn, Theo, Frederick S. vom Saal, and Ana M. Soto. 1993. "Developmental Effects of Endocrine-Disrupting Chemicals in Wildlife and Humans." *Environmental Health Perspectives* 101(5): 378–384.)

TEDX has adopted three principles as the basis of its activities. First, it relies absolutely and entirely on scientific evidence in collecting and analyzing information on endocrine disruptors. Second, it uses that information as the basis for educational programs for the general public, policymakers, and other stakeholders on the effects of chemicals on endocrine function. Third, it acts as an advocate before policy-making bodies in an attempt to develop legislation, regulation, and other actions that will reflect the best information available about endocrine disruption by chemicals.

TEDX is run by a board of directors of nine members from the fields of biology, environmental health, medicine, environmental law, public health, and related fields. Its staff consists of Colborn, Executive Director Carol Kwiatkowski, and a small number of other researchers.

TEDX divides its research into four major categories: the nature of endocrine disruption by external influences, prenatal origins of endocrine disruption, chemicals used in fossil fuel extraction, and pesticides. The organization's research on fossil fuel extraction, as an example, focuses on subcategories such as the relationship of chemicals and human health, the nature and characteristics of so-called *pit chemicals* (chemicals found in the various types of storage pits found at oil and gas well sites), drilling chemicals, air pollution, and public health issues associated with drilling operations (which includes fracking). As an example, TEDX researchers have published data on the number and types of pit chemicals found at drilling sites in New Mexico and Wyoming, along with human health issues that may be associated with each such chemical. (See, e.g., http://endocrinedisruption.org/assets/media/documents/Crosby25-3well summary4-20-09Final.pdf.) A summary of TEDX's study of chemical uses in the fracking process with potential effects on human health can be found in the organization's report at http://endocrinedisruption.org/assets/media/documents/Multistates ummary1-27-11Final.pdf.

The TEDX website provides a number of very useful tools for learning more about endocrine disruptors and for teaching about the topic to college students and others. For example, it provides an interactive list of potential endocrine disruptors classified under such general categories as household products, personal care products, plastics and rubber products, pesticides, industrial additives, and solvents. In addition, a complete list of all possible endocrine disruptors identified by TEDX is available at http://endocrinedisruption.org/endocrine-disruption/tedx-list-of-potential-endocrine-disruptors/chemicalsearch?action=search&sall=1. Another useful interactive resource is the TEDX Critical Windows of Development tool, which outlines in detail the steps that occur in the developing fetus and embryo and the types of chemicals that may interfere with normal development at each stage of growth. (See http://endocrinedisruption.org/prenatal-origins-of-endocrine-disruption/critical-windows-of-development/timeline-test/.)

Tom L. Ward (1959–)

Ward is one of a pair of the most famous oil and gas men working in the United States in the last quarter of the twentieth century and the first two decades of the twenty-first century. In 1989, he cofounded the Chesapeake Energy company with fellow Oklahoman Aubrey McClendon. The company specialized in the purchase and leasing of land for the purpose of drilling for oil and gas in the 1980s and 1990s and eventually reached the pinnacle of their business with the company becoming the largest or second largest producer of natural gas by the early 2000s.

Tom L. Ward was born on July 11, 1959, just three days before his partner-to-be McClendon, in Shattuck, Oklahoma. He later moved with his family to nearby Seiling, where he attended elementary and high schools. His family life was not a particularly easy one as his father had a tendency to drink too heavily and died at the age of 48. Although he was not particularly interested in academics, Ward decided to continue his education at the University of Oklahoma, where he majored in petroleum land management. He graduated with a bachelor of business administration in that field in 1981. After earning his degree, Ward joined a land brokerage firm owned by his uncle as a landman. A landman in the petroleum industry is a person who arranges for the purchase or leasing of land, or of mineral rights under a person's land, for the purpose of drilling for oil and/or gas.

At the time, Aubrey McClendon was a competitor of Ward's in this regard, a landman who was knocking on the same doors as was Ward to find land that could be used for petroleum exploration. As the two men got to know each other better, they decided to join forces and create a new firm. Undecided as to exactly what to call the firm, they eventually settled on Chesapeake Energy, after McClendon's favorite vacation region. At first the company was underfunded and understaffed, but it did not lack in enthusiasm and a willingness to try new technologies to look for oil and gas in regions that other companies had ignored or abandoned as holding little promise. Over a period of 20 years, Chesapeake prospered to an extent that no one could have predicted in 1989 and eventually became one of the two leading producers of natural gas in the United States.

In 2006, Ward decided to leave Chesapeake and strike out on his own. He paid $500 million for a controlling interest in Riata Energy, changed the company's name to SandRidge Energy, and became chairman and CEO of the company. In 2013, a dispute arose over Ward's right to profit from private land deals with possible conflicts with SandRidge interests, and he was dismissed from his post as chairman and CEO. He was later found innocent of any wrong dealing with regard to this issue but, by then, had moved on to form a new company, Tapstone Energy, where he is currently the CEO.

In 2000, Ward and his son Trent founded a home for neglected and abused boys in Piedmont, Oklahoma, called White Fields. The home provides a living situation and educational experience for boys until they reach high school graduation age. The White Fields program emphasizes experiences that will help boys recover from the traumatic experiences in their lives by providing training in communication skills, crisis management skills, and behavior management skills. Ward is also on the boards of Anderson University in Anderson, Indiana, and The First Tee, an educational program that aims to teach young boys and girls essential life skills through the game of golf. He is also a member of the Economic Advisory Council of the Federal Reserve Bank of Kansas City and the Board of Visitors for the Oklahoma University Health Sciences Center Department of Medicine. Until early 2014, Ward was also a co-owner of the Oklahoma City Thunder of the National Professional Basketball League.

Waterkeeper Alliance

17 Battery Pl., Suite 1329
New York, NY 10004
Phone: (212) 747-0622
Fax: (212) 747-0611
URL: http://waterkeeper.org/
Email: info@waterkeeper.org

The origins of the Waterkeeper Alliance date to the mid-1960s when a former writer for Sports Illustrated magazine, Robert Boyle, became concerned about the health of the Hudson River near Garrison, New York, where he and many fellow fishermen and fisherwomen spent many happy hours. He decided to take action against businesses, communities, and individuals who had decided that they could use the river as their own private dumping ground. He and a number of his fellow fishing hobbyists formed an organization called the Hudson River Fisherman's Association (HRFA) to get governmental bodies to enforce antipollution laws that had been on the books for years, but had not been rigorously enforced, as well as to lobby for additional laws and regulations designed to protect the river. Before long, fishing advocates along other parts of the Hudson formed similar associations in places such as Yonkers, Manhattan, and New Jersey.

HRFA won its first environmental battle in 1968 when it sued Penn Central Pipe under two nineteenth century laws, the Rivers and Harbors Act of 1888 and the Refuse Act of 1899. It won a judgment of $2,000 against Penn Central Pipe but, more importantly, it was confirmed in its philosophy that the best way to fight for the river was to sue polluters. Before long, HRFA had won judgments against much larger firms, such as American Cyanamid, Anaconda Wire and Copper, Ciba-Geigy pharmaceuticals, Standard Brands, and Westchester (New York) County. With the money earned from these judgments, the association purchased a small boat for use in patrolling the river, looking for violations of law and regulations regarding pollution of the river. In so doing, HRFA established a model by which it has continued to function to the present day, as well as one that forms the basis of the work carried out in many other types of waterways by the Waterkeeper Alliance. By 1983, HRFA was also able to hire its first full-time riverkeeper, John Cronin, whose job was to patrol the river on the association's boat.

As word of HRFA's success spread, similar organizations began to spring up across the United States and around the world. These organizations focused not only on rivers, but also on bays, bayous, canals, channels, coasts, creeks, deltas, gulfs, lakes, sounds, and other bodies of water. By 1999, there was enough interest in the concept of a waterkeeper that a number of local organizations banded together to form the Waterkeeper Alliance. That organization today claims to be the fastest growing environmental organization in the world with more than 200 separate organizations monitoring virtually every conceivable type of waterway worldwide. Examples of the organizations that are members of the Waterkeeper Alliance are the Detroit Riverkeeper, St. Clair Channelkeeper, California Coastkeeper Alliance, Puget Soundkeeper Alliance,

Ogeechee Riverkeeper, Savannah Riverkeeper, Hann (Senegal) Baykeeper, Himalayan Glacier Waterkeeper, Upper Yellow River Waterkeeper, and Punta Abreojos (Mexico) Coastkeeper.

Waterkeeper Alliance is run by a board of directors of 14 representatives from local waterkeeper group and affiliated organizations, chaired by Robert F. Kennedy, Jr., currently president of the association. In addition to local waterkeeper groups, the organization is affiliated with a number of like-minded organizations and associations, such as Toyota, Levi Strauss, Fonseca, Paul Mitchell, Teva, Patagonia, and Omega. Board members also represent a number of associated business, such as AbTech Industries, Coolglobes, Cedar Green Homes, and Seventh Generation. The organization publishes a biannual magazine, *Waterkeeper*, which is available online at http://waterkeeper.org/waterkeeper-magazine/.

Waterkeeper Alliance organizes its activities around three major themes: clean water defense; pure farms, pure waters; and clean and safe energy. The first theme follows in the long tradition of waterkeepers, the effort to make sure that rivers, lakes, and other waterways are kept clean of pollution caused by industrial, municipal, and other sources. In general, laws are now generally in place to prevent such pollution, so that the waterkeeper's job is to find instance of violation of those laws and then to instigate actions to correct that problem. The pure farms, pure waters initiative is aimed specifically at the types of pollution produced by modern agricultural, husbandry, and dairy practices, in which very large quantities of plant and animal waste, as well as a host of synthetic chemicals, are released into the natural environment, with potential harm to waterways. The third theme is concerned with the potential for damage to waterways from a variety of mining and other extraction technologies designed to produce coal, oil, and natural gas. In this regard, waterkeepers have taken a stand against fracking because of the serious pollution problems they believe that practice causes to natural water supplies. The Waterkeeper Alliance policy on fracking is that the organization will oppose shale gas extraction by hydrofracking "unless, and until, the industry proves it can be done safely to protect human health and the environment."

DON'T FRACK OUR FUTURE FOR GREED

100% WELL CASINGS FAIL OVER 100 YEARS

-Kill HB 2615-
BAN FRACKING!

If terrorists poisoned our water...

Introduction

One means of gaining an insight into the history, current status, and controversies associated with hydraulic fracturing is to read through some of the laws, regulations, court cases, speeches, reports, and other documents that have been produced on these topics. A review of relevant data and statistics also provides information on these subjects. This chapter provides excerpts from some of the most important documents dealing with fracking issues as well as basic data and statistics and the use and effects of the procedure.

Data

One of the most important reports about hydraulic fracturing is one ordered by the U.S. Congress in its FY2010 Appropriations Committee Conference Report. That report required the EPA to "study the relationship between hydraulic fracturing and drinking water, using the best available science, independent sources of information, and to conduct the study in consultation with others using a transparent, peer-reviewed

---

Environmental group members opposing oil drilling show support inside the Capitol rotunda in an effort to pressure lawmakers for a two-year moratorium on fracking, at the Illinois State Capitol in Springfield. (AP Photo/Seth Perlman)

process." In 2014, the EPA issued a Progress Report that included data that had been collected thus far. Table 5.1 summarizes data on potentially hazardous chemicals present

Table 5.1 Chemicals Present in 652 Different Products Used by Hydraulic Fracturing Companies

Chemical	Category[1]	Number of Products
Methanol	HAP	342
Ethylene glycol	HAP	119
Naphthalene	HPA	44
Xylene	SCWA	44
Hydrochloric acid	HAP	42
Toluene	SCWA, HAP	29
Ethylbenzene	SDWA, HAP	28
Diethanolamine	HAP	14
Formaldehyde	Carcinogen, HAP	12
Thiourea	Carcinogen	9
Benzyl chloride	Carcinogen, HAP	8
Cumene	HPA	6
Nitrilotriacetic acid	Carcinogen	6
Dimethylformamide	HAP	5
Phenol	HAP	5
Benzene	Carcinogen, HAP	3
Di(2-ethylhexyl)phthalate	Carcinogen, SDWA, HAP	3
Acrylamide	Carcinogen, SDWA, HAP	2
Hydrofluoric acid	HAP	2
Phthalic anhydride	HAP	2
Acetaldehyde	Carcinogen, HAP	1
Acetophenone	HAP	1
Copper	SCWA	1
Ethylene oxide	Carcinogen, HAP	1
Lead	Carcinogen, SDWA, HAP	1
Propylene oxide	Carcinogen, HAP	1
p-xylene	HAP	1

[1]HAP = Hazardous air pollutant.
SDWA = Regulated under the Safe Drinking Water Act.

Source: "Study of the Potential Impacts of Hydraulic Fracturing on Drinking Water Resources. Progress Report," Washington, DC: U. S. Environmental Protection Agency, Office of Research and Development. December 2012. EPA/601/R-12/011, Table 11, page 29. Available online at http://www2.epa.gov/sites/production/files/documents/hf-report20121214.pdf#page=209. Accessed on May 26, 2014.

in fracking products used by 11 companies who provided that service.

In 2011, the U.S. Energy Information Administration conducted a study trying to determine the amount of shale gas and shale oil reserves available in the United States. Table 5.2 provides a summary of the findings of that survey.

Table 5.2 Discovered but Unproved Technically Recoverable Shale Gas and Oil Resources in the United States: Shale Oil Resources

Play	Estimated Resource[1]	Area[2]		Expected Ultimate Recovery[3]
		Leased	Unleased	
Total Northeast	**472.05**	**101,655**	**128,272**	**0.74**
Marcellus	410.34	10,622	84,271	1.18
Big Sandy	7.40	8,675	1,994	0.33
Low Thermal Maturity	13.53	45,844		0.30
Greater Siltstone	8.46	22,914		0.19
New Albany	10.95	1,600	41,900	1.10
Antrim	19.93	12,000		0.28
Cincinnati Arch	1.44	Not Assessed		
Gulf Coast	**99.99**	**7,093**	**5,426**	**2.99**
Haynesville	74.71	3,574	5,426	3.57
Eagle Ford	20.81	1,090		5.00
FloydNeal & Conasauga	4.37	2,429		0.90
Mid-Continent	**59.88**	**14,388**		**2.45**
Fayetteville	31.96	9,000		2.07
Woodford	22.21	4,70		2.98
Cana Woodford	5.72	688		5.20
Southwest	**75.52**	**6,766**	**2,383**	**1.85**
Barnett	43.38	4,075	2,383	1.42
Barnett Woodford	32.15	2,691		3.07
Rocky Mountain	**43.03**	**30,511**		**0.69**
Hilliard-Baxter-Mancos	3.77	16,416		0.18
Lewis	11.63	7,506		1.30
Williston-Shallow Niobraran	6.61	Not Assessed		0.45
Mancos	21.02	6,589		1.00
Total Lower 48 States	**750.38**	**160,413**	**136,081**	**1.02**

[1]Trillion cubic feet.
[2]Square miles.
[3]Billion cubic feet per well.

(continued)

Table 5.2 (*continued*)

Play	Estimated Resource[1]	Area[2] Leased	Area[2] Unleased	Expected Ultimate Recovery[3]
Eagle Ford (Gulf Coast)	3.35	3,323		300
Avalon & Bone Springs (Southwest)	1.58	1,313		300
Bakken (Rocky Mountains)	3.59	6,522		550
Monterey/Santos (West Coast)	15.42	1,752		550
Total Lower 48 States	**23.94**	**12,910**		**460**

[1]Billions of barrels.
[2]Square miles.
[3]Thousands of barrels per well.

Source: "Review of Emerging Resources: U.S. Shale Gas and Shale Oil Plays," U.S. Energy Information Administration. July 2011. http://www.eia.gov/analysis/studies/usshalegas/pdf/usshaleplays.pdf. Accessed on May 27, 2014.

For a map showing the location of each of the plays listed in Table 5.3, see the reference listed above, Figure 1, page 6.

Table 5.3 Estimated Number of Fracking Wells in the United States

State	Fracking Wells since 2005	Fracking Wells Drilled in 2012
Arkansas	4,910	719
Colorado	18,168	1,896
Kansas	407	236
Louisiana	2,327	139
Mississippi	9	Not Available
Montana	264	174
New Mexico	1,353	482
North Dakota	5,166	1,713
Ohio	334	234
Oklahoma	2,694	Not Available
Pennsylvania	6,651	1,349
Tennessee	30	Not Available
Texas	33,753	13,540
Utah	1,336	765
Virginia	95	1
West Virginia[1]	3,275	610
Wyoming	1,126	468
Total	**81,898**	**22,326**

[1]Data for West Virginia is for permitted fracking wells, not wells that have been drilled. Data were not available on drilled wells.

Source: Ridlington, Elizabeth, and John Rumpler. 2013. "Fracking by the Numbers: Key Impacts of Dirty Drilling at the State and National Level," Boston: Environment America, Table 1, page 20. Available online at http://www.environmentamerica.org/sites/environment/files/reports/EA_FrackingNumbers_scrn.pdf. Accessed on May 26, 2014.

Table 5.4 Water Used for Fracking

State	Water Used for Fracking since 2005[1]
Arkansas	26,000
Colorado	26,000
Kansas	670
Louisiana	12,000
Mississippi	64
Montana	450
New Mexico	1,300
North Dakota	12,000
Ohio	1,400
Oklahoma	10,000
Pennsylvania	30,000
Tennessee	130
Texas	110,000
Utah	590
Virginia	15
West Virginia	17,000
Wyoming	1,200
Total	**250,000**

[1] Millions of gallons.

Source: Ridlington, Elizabeth, and John Rumpler. 2013. "Fracking by the Numbers: Key Impacts of Dirty Drilling at the State and National Level," Boston: Environment America, Table 3, page 23. Available online at http://www.environmentamerica.org/sites/environment/files/reports/EA_FrackingNumbers_scrn.pdf. Accessed on May 26, 2014.

In October 2013, the organization Environment America released a report on the impact of hydraulic fracturing on the natural environment and human health. The report contains some of the most useful statistical information about the effects of fracking currently available. These Tables 5.4 and 5.5 reproduce some of those data and statistics.

Table 5.5 Estimated Air Pollution Produced from Early Stages of Fracking (Drilling and Well Completion) in 2012 (tons)

State	Particulate Matter	Nitrogen Oxides	Carbon Monoxide	VOCs[1]	Sulfur Dioxide
Arkansas	400	5,300	8,100	700	20
Colorado	1,100	14,000	21,000	2,000	50

(continued)

Table 5.5 (*continued*)

State	Particulate Matter	Nitrogen Oxides	Carbon Monoxide	VOCs[1]	Sulfur Dioxide
Kansas	100	1,700	2,700	200	6
Louisiana	80	1,000	1,600	100	3
Mississippi			Data Unavailable		
Montana	100	1,300	2,000	200	4
New Mexico	300	3,600	5,400	500	10
North Dakota	1,000	13,000	19,000	2,000	40
Ohio	100	1,700	2,600	200	6
Oklahoma			Data Unavailable		
Pennsylvania	800	10,000	15,000	1,000	30
Tennessee			Data Unavailable		
Texas	7,800	100,000	153,000	14,000	300
Utah	400	5,700	9,000	1,000	20
Virginia	1	7	11	1	0
West Virginia	400	4,500	6,900	600	20
Wyoming	270	3,500	5,300	500	12
Total	**13,000**	**170,000**	**250,000**	**23,000**	**600**

[1]Volatile organic compounds.

Source: Ridlington, Elizabeth, and John Rumpler. 2013. "Fracking by the Numbers: Key Impacts of Dirty Drilling at the State and National Level," Boston: Environment America, Table 4, page 23. Available online at http://www.environmentamerica.org/sites/environment/files/reports/EA_FrackingNumbers_scrn.pdf. Accessed on May 26, 2014.

The U.S. Energy Information Administration periodically studies shale oil and gas reserves available in a number of nations around the world. The most recent study was reported in 2013, and its general results are summarized in Table 5.6.

Table 5.6 Estimated Risked Shale Gas and Shale Oil Resources In-Place and Technically Recoverable in 41 Countries as Assessed in 2013

Country/Area	Risked Gas in Place (Tcf)[1]	Technically Recoverable Gas (Tcf)[1]	Risked Oil in Place (billion bbl)[2]	Technically Recoverable Oil (billion bbl)[2]
United States[3]	**4,644**	**1,161**	**954**	**47.7**
Canada	2,413	573	162	8.8
Mexico	2,233	545	275	13.1

(*continued*)

Table 5.6 *(continued)*

Country/Area	Risked Gas in Place (Tcf)[1]	Technically Recoverable Gas (Tcf)[1]	Risked Oil in Place (billion bbl)[2]	Technically Recoverable Oil (billion bbl)[2]
North America[4]	**4,647**	**1,118**	**437**	**21.9**
Australia	**2,046**	**437**	**403**	**17.5**
Colombia	308	55	120	6.8
Venezuela	815	167	269	13.4
Argentina	3,244	802	480	27.0
Brazil	1,279	245	134	5.3
Bolivia	154	36	11	0.6
Chile	228	48	47	2.3
Paraguay	350	75	77	3.7
Uruguay	13	2	14	0.6
South America	**6,390**	**1,431**	**1,152**	**59.7**
Poland	763	148	65	3.3
Lithuania	4	0	5	0.3
Kaliningrad	20	2	24	1.2
Russia	1,921	285	1,243	74.6
Bulgaria	66	17	4	0.2
Romania	233	51	6	0.3
Ukraine	572	128	23	1.1
United Kingdom	134	26	17	0.7
Spain	42	8	3	0.1
France	727	137	118	4.7
Germany	80	17	14	0.7
Netherlands	151	26	59	2.9
Denmark	159	32	0	0.0
Sweden	49	10	0	0.0
Europe	**4,895**	**883**	**1,551**	**88.6**
Morocco	95	20	5	0.2
Algeria	3,419	707	121	5.7
Tunisia	114	23	29	1.5
Libya	942	122	613	26.1
Egypt	535	100	114	4.6
South Africa	1,559	390	0	0.0
Africa	**6,664**	**1,361**	**882**	**38.1**
China	4,746	1,115	644	32.2
Mongolia	55	4	85	3.4
Thailand	22	5	0	0.0
Indonesia	303	46	234	7.9

(continued)

Table 5.6 (*continued*)

Country/Area	Risked Gas in Place (Tcf)[1]	Technically Recoverable Gas (Tcf)[1]	Risked Oil in Place (billion bbl)[2]	Technically Recoverable Oil (billion bbl)[2]
India	584	96	87	3.8
Pakistan	586	105	227	9.1
Jordan	35	7	4	0.1
Turkey	163	24	94	4.7
Asia	**6,495**	**1,403**	**1,375**	**61.1**
World	**31,138**	**6,634**	**5,799**	**286.9**

[1]Trillion cubic feet.
[2]Billion barrels.
[3]For purposes of comparison only.
[4]Excluding the United States.

Source: "Technically Recoverable Shale Oil and Shale Gas Resources: An Assessment of 137 Shale Formations in 41 Countries Outside the United States," U.S. Energy Information Administration. June 2013. http://www.eia.gov/analysis/studies/worldshalegas/pdf/fullreport.pdf. Accessed on May 31, 2014.

Table 5.7 presents a summary of crude oil and natural gas petroleum reserves in the United States from 1900 to the present day, as determined by the U.S. Energy Information Administration.

Table 5.7 U.S. Crude Oil and Natural Gas Proved Reserves[1]

Year	Crude Oil	Natural Gas[2]
1900	2,900	
1910	4,500	
1920	7,200	
1930	13,600	
1940	19,025	
1950	25,268	
1960	31,613	
1970	39,001	
1980	29,805	206,259
1990	26,254	177,756
2000	22,045	186,510
2005	21,757	213,308
2006	20,972	220,416

(*continued*)

Table 5.7 *(continued)*

Year	Crude Oil	Natural Gas[2]
2007	21,317	247,789
2008	19,121	255,035
2009	20,682	283,879
2010	23,267	317,647
2011	26,544	348,809
2012	30,529	322,670

[1]Million barrels of oil and billion cubic feet of gas.
[2]Data not available prior to 1979.

Source: "U.S. Crude Oil Proved Reserves," U.S. Energy Information Administration. April 10, 2014. http://www.eia.gov/dnav/pet/hist/LeafHandler.ashx?n=PET&s=RCRR01NUS_1&f=A. Accessed on July 7, 2014; "U.S. Natural Gas, Wet after Lease Separation Proved Reserves." U.S. Energy Information Administration. April 10, 2014. http://www.eia.gov/dnav/ng/hist/rngr21nus_1a.htm. Accessed on July 7, 2014.

Documents

Crude Oil Windfall Profit Tax Act of 1980

One of the earliest efforts by the federal government to encourage the development of alternative sources of energy was the Crude Oil Windfall Profit Tax Act of 1980, which provided tax incentives for a number of such energy sources. The section excerpted here deals with natural gas and oil extracted from Devonian shale, a technology later greatly aided and developed by the use of hydraulic fracturing.

PART III—PRODUCTION OF FUEL FROM NONCONVENTIONAL SOURCES; ALCOHOL FUELS

SEC. 231. PRODUCTION TAX CREDIT.

(a) IN GENERAL.—Subpart A of part IV of subchapter A of chapter 1 (relating to credits against tax) is amended by inserting after section 44C the following new section:

"SEC. 44D. CREDIT FOR PRODUCING FUEL FROM A NONCONVENTIONAL SOURCE.

"(a) ALLOWANCE OF CREDIT.—There shall be allowed as a credit against the tax imposed by this chapter for the taxable year an amount equal to—

"(1) $3, multiplied by

"(2) the barrel-of-oil equivalent of qualified fuels—

"(A) sold by the taxpayer to an unrelated person during the taxable year,
and

"(B) the production of which is attributable to the taxpayer.

. . .

"(c) DEFINITION OF QUALIFIED FUELS—For purposes of this section—

"(1) IN GENERAL.—The term 'qualified fuels' means—

"(A) oil produced from shale and tar sands,

"(B) gas produced from—

"(i) geopressured brine, Devonian shale, coal seams, or
a tight formation, or

"(ii) biomass,

"(C) liquid, gaseous, or solid synthetic fuels produced from coal (including lignite), including such fuels when used as feedstocks,

"(D) qualifying processed wood fuels, and

"(E) steam produced from solid agricultural byproducts (not including timber byproducts).

"(2) GAS FROM GEOPRESSURED BRINE, ETC.—

"(A) IN GENERAL.—Except as provided in subparagraph (B), the determination of whether any gas is produced from geopressured brine, Devonian shale, coal seams, or a tight formation shall be made in accordance with section 503 of the Natural Gas Policy Act of 1978.

"(B) SPECIAL RULES FOR GAS FROM TIGHT FORMATIONS.—The term 'gas produced from a tight formation' shall only include—

"(i) gas the price of which is regulated by the United States, and

"(ii) gas for which the maximum lawful price applicable under the Natural Gas Policy Act of 1978 is at least 150 percent of the then applicable price under section 103 of such Act.

. . .

(4) SPECIAL RULES APPUCABLE TO GAS FROM GEOPRESSURED BRINE, DEVONIAN SHALE, COAL SEAMS, OR A TIGHT FORMATION.—

"(A) CREDIT ALLOWED ONLY FOR NEW PRODUCTION.—The amount of the credit allowable under subsection (a) shall be determined without regard to any production attributable to a property from which gas from Devonian shale, coal seams, geopressured brine, or a tight formation was produced in marketable quantities before January 1, 1980.

"(B) REFERENCE PRICE AND APPUCATION OF PHASEOUT FOR DEVONIAN SHALE.—

"(i) REFERENCE PRICE FOR DEVONIAN SHALE—For purposes of this section, the term 'reference price' for gas from Devonian shale sold during calendar years 1980, 1981, and 1982 shall be the average wellhead price per thousand cubic feet for such year of high cost natural gas (as defined in section 107 (c) (2), (3), and (4) of the Natural Gas Policy Act of 1978 and determined under section 503 of that act) as estimated by the Secretary after consultation with the Federal Energy Regulatory Commission.

"(ii) DIFFERENT PHASEOUT TO APPLY FOR 1980, 1981, AND 1982.—For purposes of applying paragraphs (1) and (2) of subsection (b) with respect to sales during calendar years 1980, 1981, and 1982 of gas from Devonian shale, '$4.05' shall be substituted for '$23.50' and '$1.03' shall be substituted for '$6.00'.

Source: Public Law 96-223—April 2, 1980. http://www.gpo.gov/fdsys/pkg/STATUTE-94/pdf/STATUTE-94-Pg229.pdf. Accessed on May 27, 2014.

Hydraulic Fracturing Exclusions

The process of hydraulic fracturing has been specifically excluded from provisions of a number of federal laws originally written to protect the natural environment and human health. Examples of some of these exceptions provided to the fracking process are provided here.

Safe Drinking Water Act, Public Law 113-103, Section 300h(d)(1)

(d) "Underground injection" defined; underground injection endangerment of drinking water sources

For purposes of this part:

(1) Underground injection.—The term "underground injection"—
 (A) means the subsurface emplacement of fluids by well injection; and
 (B) excludes—
  (i) the underground injection of natural gas for purposes of storage; and
  (ii) the underground injection of fluids or propping agents (other than diesel fuels) pursuant to hydraulic fracturing operations related to oil, gas, or geothermal production activities.

(2) Underground injection endangers drinking water sources if such injection may result in the presence in underground water which supplies or can reasonably be expected to supply any public water system of any contaminant, and if the presence of such contaminant may result in such system's not complying with any national primary drinking water regulation or may otherwise adversely affect the health of persons.

Source: U.S. Code, Title 42, Section 300h(d)(1). http://uscode.house.gov/download/download.shtml. Accessed on May 28, 2014.

Clean Water Act, Public Law 95-217, Sections 1326, 1342

(2) Stormwater runoff from oil, gas, and mining operations The Administrator shall not require a permit under this section, nor shall the Administrator directly or indirectly require any State to require a permit, for discharges of stormwater runoff from mining operations or oil and gas exploration, production, processing, or treatment operations or transmission facilities, composed entirely of flows which are from conveyances or systems of conveyances (including but not limited to pipes,

conduits, ditches, and channels) used for collecting and conveying precipitation runoff and which are not contaminated by contact with, or do not come into contact with, any overburden, raw material, intermediate products, finished product, byproduct, or waste products located on the site of such operations.

. . .

(6) The term "pollutant" means dredged spoil, solid waste, incinerator residue, sewage, garbage, sewage sludge, munitions, chemical wastes, biological materials, radioactive materials, heat, wrecked or discarded equipment, rock, sand, cellar dirt and industrial, municipal, and agricultural waste discharged into water. This term does not mean (A) "sewage from vessels or a discharge incidental to the normal operation of a vessel of the Armed Forces" within the meaning of section 1322 of this title; or (B) water, gas, or other material which is injected into a well to facilitate production of oil or gas, or water derived in association with oil or gas production and disposed of in a well, if the well used either to facilitate production or for disposal purposes is approved by authority of the State in which the well is located, and if such State determines that such injection or disposal will not result in the degradation of ground or surface water resources.

Source: U.S. Code, Title 33, Sections 1326(24) and 1342(l)(2). http://uscode.house.gov/download/download.shtml. Accessed on May 28, 2014.

Resource Conservation and Recovery Act (RCRA), Public Law 94-580, 6921(b)

Notwithstanding the provisions of paragraph (1) of this subsection, drilling fluids, produced waters, and other wastes associated with the exploration, development, or production of crude oil or natural gas or geothermal energy shall be subject only to existing State or Federal regulatory programs in lieu of this subchapter until at least 24 months after October 21,

1980, and after promulgation of the regulations in accordance with subparagraphs (B) and (C) of this paragraph.

Source: U.S. Code, Title 42, Section 6921. http://uscode.house.gov/download/download.shtml. Accessed on May 28, 2014.

This law was later clarified by two EPA regulations posted in the Federal Register, the first of which listed nearly three pages of materials exempted from provisions of *the RCRA, including, as examples:*

—Produced water; Drilling fluids; Drill cuttings; Rigwash; Drilling fluids and cuttings from offshore operations disposed of onshore; Well completion, treatment, and stimulation fluids; Basic sediment and water and other tank bottoms from storage facilities that hold product and exempt waste; Accumulated materials such as hydrocarbons, solids, sand, and emulsion from production separators, fluid treating vessels, and production impoundments; Pit sludges and contaminated bottoms from storage or disposal of exempt wastes; Packing fluids; Produced sand; Pipe scale, hydrocarbon solids, hydrates, and other deposits removed from piping and equipment prior to transportation; Hydrocarbon-bearing soil; Wastes from subsurface gas storage and retrieval, except for the nonexempt wastes listed below; Constituents removed from produced water before it is injected or otherwise disposed of; Liquid hydrocarbons removed from the production stream but not from oil refining; Gases from the production stream, such as hydrogen sulfide and carbon dioxide, and volatilized hydrocarbons; Materials ejected from a producing well during the process known as blowdown; Waste crude oil from primary field operations and production; Light organics volatilized from exempt wastes in reserve pits or impoundments or production equipment; Unused fracturing fluids or acids; Painting wastes; Oil and gas service company wastes, such as empty drums, drum rinsate, vacuum truck rinsate, sandblast media, painting wastes, spent solvents, spilled chemicals, and waste acids; Used equipment lubrication oils; Used hydraulic fluids; Waste solvents.

Source: Regulatory Determination for Oil and Gas and Geothermal Exploration, Development and Production Wastes. 53 Fed. Reg. 25,447. http://www.epa.gov/osw/nonhaz/industrial/special/oil/og88wp.pdf. Accessed on May 28, 2014. See also Clarification of the Regulatory Determination for Wastes from the Exploration, Development, and Production of Crude Oil, Natural Gas and Geothermal Energy. 58 Fed. Reg. 15,284. http://www.epa.gov/osw/nonhaz/industrial/special/oil/og93wp.pdf. Accessed on May 28, 2014.

Comprehensive Environmental Response, Compensation, and Liability Act (CERCLA; also known as the Superfund Act), Public Law 96-510, 9601

The so-called Superfund Act contains a number of sections that exempt oil and gas industries from provisions of the act. One such section involves the definition of hazardous substances under the act.

(14) The term "hazardous substance" means . . .

(F) any imminently hazardous chemical substance or mixture with respect to which the Administrator has taken action pursuant to section 7 of the Toxic Substances Control Act [15 U.S.C. 2606]. The term does not include petroleum, including crude oil or any fraction thereof which is not otherwise specifically listed or designated as a hazardous substance under subparagraphs (A) through (F) of this paragraph, and the term does not include natural gas, natural gas liquids, liquefied natural gas, or synthetic gas usable for fuel (or mixtures of natural gas and such synthetic gas).

. . .

(33) The term "pollutant or contaminant" shall include, but not be limited to, any element, substance, compound, or

mixture, including disease-causing agents, which after release into the environment and upon exposure, ingestion, inhalation, or assimilation into any organism, either directly from the environment or indirectly by ingestion through food chains, will or may reasonably be anticipated to cause death, disease, behavioral abnormalities, cancer, genetic mutation, physiological malfunctions (including malfunctions in reproduction) or physical deformations, in such organisms or their offspring; except that the term "pollutant or contaminant" shall not include petroleum, including crude oil or any fraction thereof which is not otherwise specifically listed or designated as a hazardous substance under subparagraphs (A) through (F) of paragraph (14) and shall not include natural gas, liquefied natural gas, or synthetic gas of pipeline quality (or mixtures of natural gas and such synthetic gas).

Source: U.S. Code, Title 42, Section 9601. http://uscode.house.gov/download/download.shtml. Accessed on May 28, 2014.

Legal Environmental Assistance Foundation vs. U.S. EPA, 118 F.3d 1467 (1997)

The question as to whether or not the U.S. Environmental Protection Agency (EPA) could or must regulate hydraulic fracturing has a long history, beginning with an appeal by the Legal Environmental Assistance Foundation (LEAF) to the EPA in 1994 that they begin doing just that. When the EPA declined to do so, LEAF filed suit, which eventually reached the 11th Circuit Court of Appeals. That court's ruling in this matter was as follows. (Ellipses indicate the deletion of references and/or citations.)

Contrary to EPA, "[w]e do not start from the premise that [the statutory] language is imprecise. Instead, we assume that in drafting legislation, Congress said what it meant." . . . It is only after we have determined that words used by Congress are ambiguous, or that Congress left a gap in the statutory language, that we turn to the agency's interpretation of these words to ascertain whether it deserves any deference. … "Giving the words used their ordinary meaning," … we readily find that the word "injection" means the act of "forc[ing] (a fluid) into a passage, cavity, or tissue." … Sensibly, therefore, "underground injection" means the subsurface emplacement of fluids by forcing them into cavities and passages in the ground through a well. … The process of hydraulic fracturing obviously falls within this definition, as it involves the subsurface emplacement … of fluids by forcing them into cracks in the ground through a well. Nothing in the statutory definition suggests that EPA has the authority to exclude from the reach of the regulations an activity (i.e., hydraulic fracturing) which unquestionably falls within the plain meaning of the definition, on the basis that the well that is used to achieve that activity is also used—even primarily used for another activity (i.e., methane gas production) that does not constitute underground injection. EPA's argument that a methane gas production well is not an "injection well" because it is used primarily for gas extraction is spurious. Congress directed EPA to regulate "underground injection" activities, not "injection wells." In view of clear statutory language requiring the regulation of all such activities, they must be regulated, regardless of the other uses of the well in which these activities occur. …

Source: *Legal Environmental Assistance Foundation vs. U.S. EPA*, 118 F.3d 1467. https://law.resource.org/pub/us/case/reporter/F3/118/118.F3d.1467.95-6501.html. Accessed on May 25, 2014.

Evaluation of Impacts to Underground Sources of Drinking Water by Hydraulic Fracturing of Coalbed Methane Reservoirs Study (2004)

In 2002, the EPA initiated a study on the possible effects of hydraulic fracturing operations on the quality of drinking water in regions where such operations were being conducted. The agency released its final report on this topic in June 2004. The study found that fracking has no significant effects on drinking water quality, a finding that was soon to be challenged by other individuals and agencies. One complaint raised by these critics was that the EPA had been too strongly influenced by the pro-industry bent of President George W. Bush and his vice president (and former official at the Halliburton Company), Richard Cheney. The report's major conclusion in this regard was as follows:

[From the executive summary of the report:]

Based on the information collected and reviewed, EPA has concluded that the injection of hydraulic fracturing fluids into CBM wells poses little or no threat to USDWs and does not justify additional study at this time.

[From a more detailed discussion of this point later in the report:]

ES-8 Did EPA Find Any Cases of Contaminated Drinking Water Wells Caused by Hydraulic Fracturing in CBM Wells?

EPA did not find confirmed evidence that drinking water wells have been contaminated by hydraulic fracturing fluid injection into CBM wells. EPA reviewed studies and follow-up investigations conducted by state agencies in response to citizen reports that CBM production resulted in water quality and quantity incidents. In addition, EPA received reports from concerned citizens in each area with significant CBM development.

These complaints pertained to the following basins:

- San Juan Basin (Colorado and New Mexico);
- Powder River Basin (Wyoming and Montana);
- Black Warrior Basin (Alabama); and
- Central Appalachian Basin (Virginia and West Virginia).

Examples of concerns and claims raised by citizens include:

- Drinking water with strong, unpleasant taste and odor.
- Impacts on fish, and surrounding vegetation and wildlife
- Loss of water in wells and aquifers, and discharged water creating artificial ponds and swamps not indigenous to region.

Water quantity complaints were the most predominant cause for complaint by private well owners. After reviewing data and incident reports provided by states, EPA sees no conclusive evidence that water quality degradation in USDWs is a direct result of injection of hydraulic fracturing fluids into CBM wells and subsequent underground movement of these fluids. Several other factors may contribute to groundwater problems, such as various aspects of resource development, naturally occurring conditions, population growth, and historical well-completion or abandonment practices.

Many of the incidents that were reported (such as water loss and impacts on nearby flora and fauna from discharge of produced water) are beyond the authorities of EPA under SDWA and the scope of Phase I of this study.

Source: Executive Summary, Evaluation of Impacts to Underground Sources of Drinking Water by Hydraulic Fracturing of Coalbed Methane Reservoirs Study. http://www.epa.gov/ogwdw/uic/pdfs/cbmstudy_attach_uic_exec_summ.pdf. Accessed on May 30, 2014.

Testimony Submitted to the House Committee on Natural Resources Subcommittee on Energy and Mineral Resources Washington, DC, June 18, 2009, Prepared by the Interstate Oil and Gas Compact Commission on Behalf of the Nation's Oil and Gas Producing States

In this presentation, the Interstate Oil and Gas Compact Commission (IOGCC) argues that fracking is a common and safe practice in the extraction of oil and gas in the United States. The presentation is of value to the general reader largely because of the five attachments offered as support for the IOGCC's argument that studies have shown that fracking causes no harm to drinking water with documents from the federal law, relevant court cases, reports from state oil and gas commissions, resolutions from state legislative bodies, and previous testimony from an IOGCC representative. The case made in the presentation is summarized in the initial portion of the document, as follows:

Summary:

Hydraulic fracturing has been used safely to stimulate oil and gas production in the United States for more than 60 years and is thoroughly regulated by the oil and gas regulatory agencies of the member states of the Interstate Oil and Gas Compact Commission (IOGCC).

Additional study is unnecessary, and in fact, would be a wasteful use of taxpayers' dollars. However, all future studies involving the regulation of oil and natural gas exploration and production must involve leadership by those officials who know it best—state regulators.

Claims that hydraulic fracturing has been linked directly to the contamination of underground sources of drinking water are untrue. If factual information exists to the contrary, the public, media and policymakers are urged to contact the appropriate state officials for further investigation.

Legislators and regulators did not intend to regulate the short-term process of well stimulation by hydraulic fracturing under the U.S. Environmental Protection Agency's requirements for long-term disposal of substances (underground injection control (UIC) program).

Hydraulic fracturing plays a critical role in the development of virtually all unconventional oil and natural gas resources. The technology has significantly increased domestic reserves, especially clean-burning natural gas. Further regulatory burdens are unnecessary, and in fact, would delay the development of vital domestic natural gas resources and increase energy costs to the consumer with no resulting environmental benefit.

Source: Testimony Submitted to the House Committee on Natural Resources Subcommittee on Energy and Mineral Resources. Washington, DC, June 18, 2009. Prepared by the Interstate Oil and Gas Compact Commission on Behalf of the Nation's Oil and Gas Producing States. https://iogcc.publishpath.com/Websites/iogcc/Images/Additional-IOGCC-Testimony-June 2009.pdf. Accessed on July 11, 2014.

Study of the Potential Impacts of Hydraulic Fracturing on Drinking Water Resources—Progress Report (2012)

In 2010, the U.S. Congress directed the EPA to conduct a comprehensive study of the effects of hydraulic fracturing on drinking water. Following that directive, the EPA began a long-range study on the topic, producing a progress report a few years after the study began. The executive summary of that report describes the type of work the EPA had planned to do to fulfill this mission.

Executive Summary

Natural gas plays a key role in our nation's clean energy future. The United States has vast reserves of natural gas that are commercially viable as a result of advances in horizontal drilling and hydraulic fracturing technologies, which enable greater access to gas in rock formations deep underground. These advances have spurred a significant increase in the production of both natural gas and oil across the country.

Responsible development of America's oil and gas resources offers important economic, energy security, and environmental benefits. However, as the use of hydraulic fracturing has increased, so have concerns about its potential human health and environmental impacts, especially for drinking water. In response to public concern, the US House of Representatives requested that the US Environmental Protection Agency

(EPA) conduct scientific research to examine the relationship between hydraulic fracturing and drinking water resources (*citation omitted*).

In 2011, the EPA began research under its Plan to Study the Potential Impacts of Hydraulic Fracturing on Drinking Water Resources. The purpose of the study is to assess the potential impacts of hydraulic fracturing on drinking water resources, if any, and to identify the driving factors that may affect the severity and frequency of such impacts. Scientists are focusing primarily on hydraulic fracturing of shale formations to extract natural gas, with some study of other oil- and gas-producing formations, including tight sands, and coalbeds. The EPA has designed the scope of the research around five stages of the hydraulic fracturing water cycle. Each stage of the cycle is associated with a primary research question:

- Water acquisition: What are the possible impacts of large volume water withdrawals from ground and surface waters on drinking water resources?
- Chemical mixing: What are the possible impacts of hydraulic fracturing fluid surface spills on or near well pads on drinking water resources?
- Well injection: What are the possible impacts of the injection and fracturing process on drinking water resources?
- Flowback and produced water: What are the possible impacts of flowback and produced water (collectively referred to as "hydraulic fracturing wastewater") surface spills on or near well pads on drinking water resources?
- Wastewater treatment and waste disposal: What are the possible impacts of inadequate treatment of hydraulic fracturing wastewater on drinking water resources?

This report describes 18 research projects underway to answer these research questions and presents the progress made as of September 2012 for each of the projects. Information presented as part of this report cannot be used to draw conclusions about potential impacts to drinking water resources from hydraulic fracturing. The research projects are organized according to five different types of research activities: analysis of existing data, scenario evaluations, laboratory studies, toxicity assessments, and case studies.

Source: Study of the Potential Impacts of Hydraulic Fracturing on Drinking Water Resources. Progress Report. Washington, DC: US Environmental Protection Agency, Office of Research and Development. December 2012, EPA/601/R-12/011. Available online at http://www2.epa.gov/sites/production/files/documents/hf-report20121214.pdf. Accessed on May 27, 2014.

Act 13, State of Pennsylvania (2012)

In 2012, the Pennsylvania state legislature passed a bill called Act 13 that dealt with a number of issues related to the extraction of oil and gas in the state, with special attention to unconventional sources of the fuels, such as hydraulic fracturing. Probably the most contentious feature of that act was the restriction placed by the legislature on the control that local communities were to be allowed to have over fracking operations within their boundaries. The most important provisions of the act in that regard are the following. [The controversial portion of section 3215 is section (b) (4).]

§ 3303. Oil and gas operations regulated by environmental acts.

Notwithstanding any other law to the contrary, environmental acts are of Statewide concern and, to the extent that they regulate oil and gas operations, occupy the entire field of regulation, to the exclusion of all local ordinances. The Commonwealth by this section, preempts and supersedes the local regulation of oil and gas operations regulated by the environmental acts, as provided in this chapter.

§ 3304. Uniformity of local ordinances.

(a) General rule.—In addition to the restrictions contained in sections 3302 (relating to oil and gas operations regulated pursuant to Chapter 32) and 3303 (relating to oil and gas operations regulated by environmental acts), all local ordinances regulating oil and gas operations shall allow for the reasonable development of oil and gas resources.

(b) Reasonable development of oil and gas resources.— In order to allow for the reasonable development of oil and gas resources, a local ordinance:

(1) Shall allow well and pipeline location assessment operations, including seismic operations and related activities conducted in accordance with all applicable Federal and State laws and regulations relating to the storage and use of explosives throughout every local government.

(2) May not impose conditions, requirements or limitations on the construction of oil and gas operations that are more stringent than conditions, requirements or limitations imposed on construction activities for other industrial uses within the geographic boundaries of the local government.

(3) May not impose conditions, requirements or limitations on the heights of structures, screening and fencing, lighting or noise relating to permanent oil and gas operations that are more stringent than the conditions, requirements or limitations imposed on other industrial uses or other land development within the particular zoning district where the oil and gas operations are situated within the local government.

(4) Shall have a review period for permitted uses that does not exceed 30 days for complete submissions or that does not exceed 120 days for conditional uses.

(5) Shall authorize oil and gas operations, other than activities at impoundment areas, compressor stations and processing plants, as a permitted use in all zoning districts.

. . .

(b) Limitation.—

(1) No well site may be prepared or well drilled within 100 feet or, in the case of an unconventional well, 300 feet from the vertical well bore or 100 feet from the edge of the well site, whichever is greater, measured horizontally from any solid blue

lined stream, spring or body of water as identified on the most current 7 1/2 minute topographic quadrangle map of the United States Geological Survey.

(2) The edge of the disturbed area associated with any unconventional well site must maintain a 100-foot setback from the edge of any solid blue lined stream, spring or body of water as identified on the most current 7 1/2 minute topographic quadrangle map of the United States Geological Survey.

(3) No unconventional well may be drilled within 300 feet of any wetlands greater than one acre in size, and the edge of the disturbed area of any well site must maintain a 100-foot setback from the boundary of the wetlands.

(4) The department shall waive the distance restrictions upon submission of a plan identifying additional measures, facilities or practices to be employed during well site construction, drilling and operations necessary to protect the waters of this Commonwealth.

Source: Oil and Gas (58 PA.C.S.)—Omnibus Amendments. Act of Feb. 14, 2012, P.L. 87, No. 13 Cl. 58. http://www.legis.state.pa.us/WU01/LI/LI/US/HTM/2012/0/0013.HTM. Accessed on July 7, 2014.

Robinson Township, et al. vs. Commonwealth of Pennsylvania, (J-127A-D-2012) (2013)

Act 13 (cited earlier) met with a good deal of resistance from individuals, organizations, and communities who argued that the state legislature was infringing on their personal, environmental, and zoning rights. Eventually a case bringing together seven communities with this view of the act reached the Pennsylvania Supreme Court under the title of Robinson Township, et al., vs. Commonwealth of Pennsylvania, *the "et al." being the other six communities involved in the case.*

On December 19, 2013, the court announced its decision, holding partly in support of the state and, probably most importantly, partly in favor of the plaintiffs. Critical portions of the court's decision are reprinted here.

Reviewing the amended Act, few could seriously dispute how remarkable a revolution is worked by this legislation upon the existing zoning regimen in Pennsylvania, including residential zones. In short, local government is required to authorize oil and gas operations, impoundment areas, and location assessment operations (including seismic testing and the use of explosives) as permitted uses in all zoning districts throughout a locality. Local government is also required to authorize natural gas compressor stations as permitted uses in agricultural and industrial districts, and as conditional uses in all other zoning districts. Local governments are also commanded to authorize natural gas processing plants as permitted uses in industrial districts and as conditional uses in agricultural districts. Moreover, Section 3304 limits local government to imposing conditions: on construction of oil and gas operations only as stringent as those on construction activities for industrial uses; and on heights of structures, screening and fencing, lighting and noise only as stringent as those imposed on other land development within the same zoning district. Local government is also simply prohibited from limiting subterranean operations and hours of operation for assembly and disassembly of drilling rigs, and for operation of oil and gas wells, compressor stations, or processing plants. Localities also may not increase setbacks from uses related to the oil and gas industry beyond those articulated by Act 13. In addition, the dictated approach to setbacks focuses only on "existing buildings," offering residents and property owners no setback protections should they desire to develop further their own properties. That local government's zoning role is reduced to pro forma accommodation is confirmed by the fact that review under local ordinances of proposed oil and gas-related uses must be completed in 30 days for permitted uses, and in 120 days for conditional uses. The displacement of prior planning, and derivative expectations, regarding land use, zoning, and enjoyment of property is unprecedented.

. . .

We have explained that, among other fiduciary duties under Article I, Section 27, the General Assembly has the obligation to prevent degradation, diminution, and depletion of our public natural resources, which it may satisfy by enacting legislation that adequately restrains actions of private parties likely to cause harm to protected aspects of our environment. We are constrained to hold that Section 3304 falls considerably short of meeting this obligation for two reasons.

First, a new regulatory regime permitting industrial uses as a matter of right in every type of pre-existing zoning district is incapable of conserving or maintaining the constitutionally-protected aspects of the public environment and of a certain quality of life. In Pennsylvania, terrain and natural conditions frequently differ throughout a municipality, and from municipality to municipality. As a result, the impact on the quality, quantity, and well-being of our natural resources cannot reasonably be assessed on the basis of a statewide average. Protection of environmental values, in this respect, is a quintessential local issue that must be tailored to local conditions. *[citation omitted]* Moreover, the Commonwealth is now over three centuries old, and its citizens settled the territory and built homes and communities long before the exploitation of natural gas in the Marcellus Shale Formation became economically feasible. Oil and gas operations do not function autonomously of their immediate surroundings. Act 13 emerged upon this complex background of settled habitability and ownership interests and expectations.

Despite this variety in the existing environmental and legislative landscape, Act 13 simply displaces development guidelines, guidelines which offer strict limitations on industrial uses in sensitive zoning districts; instead, Act 13 permits industrial oil and gas operations as a use "of right" in every zoning district throughout the Commonwealth, including in residential, commercial, and agricultural districts

. . .

A second difficulty arising from Section 3304's requirement that local government permit industrial uses in all zoning districts is that some properties and communities will carry much heavier environmental and habitability burdens than others. *[citation omitted]* This disparate effect is irreconcilable with the express command that the trustee will manage the corpus of the trust for the benefit of "all the people." *[citation omitted]*

. . .

[In conclusion:]

C. Sections 3215(b)(4), 3215(d), 3303, and 3304 violate the Environmental Rights Amendment. We do not reach other constitutional issues raised by the parties with respect to these provisions. As a result, the Commonwealth Court's decision is affirmed with respect to Sections 3215(b)(4) and 3304 (on different grounds), and reversed with respect to Sections 3215(d) and 3303. Accordingly, application and enforcement of Sections 3215(b)(4), 3215(d), 3303, and 3304 is hereby enjoined.

Source: Oil and Gas (58 PA.C.S.)—Omnibus Amendments. Act of Feb. 14, 2012, P.L. 87, No. 13. http://www.legis.state.pa.us/WU01/LI/LI/US/HTM/2012/0/0013..HTM. Accessed on June 2, 2014. *Robinson Township vs. Commonwealth of Pennsylvania*, (J-127A-D-2012). http://www.pacourts.us/assets/opinions/Supreme/out/J-127A-D-2012oajc.pdf?cb=1. Accessed on June 2, 2014.

Fracturing Responsibility and Awareness of Chemicals Act S.1135, 113th Congress, 1st Session (2013)

Beginning in the 111th Congress (2009), a group of legislators has introduced a bill to define hydraulic fracturing as a federally regulated activity under terms of the Safe Drinking Water Act of 1974. The bill has never made much progress through the Congress, not even having been heard in committee. It appears likely, however, that efforts by some members of Congress will continue to obtain greater federal regulation of the fracking process. Ellipses (. . .) indicate the omission of unrelated information or sources.

SEC. 2. REGULATION OF HYDRAULIC FRACTURING.

(a) Underground Injection—Section 1421(d) of the Safe Drinking Water Act (42 U.S.C. 300h(d)) is amended by striking paragraph (1) and inserting the following:

'(1) UNDERGROUND INJECTION—

(A) IN GENERAL—The term 'underground injection' means the subsurface emplacement of fluids by well injection.

'(B) INCLUSION—The term 'underground injection' includes the underground injection of fluids or propping agents pursuant to hydraulic fracturing operations relating to oil or natural gas production activities.

'(C) EXCLUSION—The term 'underground injection' does not include the underground injection of natural gas for the purpose of storage.'

(b) State Primary Enforcement Relating to Hydraulic Fracturing Operations—Section 1422 of the Safe Drinking Water Act (42 U.S.C. 300h-1) is amended by adding at the end the following:

'(f) Hydraulic Fracturing Operations—

'(1) IN GENERAL—Consistent with such regulations as the Administrator may prescribe, a State may seek primary enforcement responsibility for hydraulic fracturing operations for oil and natural gas without seeking to assume primary enforcement responsibility for other types of underground injection control wells, including underground injection control wells that inject brine or other fluids that are brought to the surface in connection with oil and natural gas production or any underground injection for the secondary or tertiary recovery of oil or natural gas.

. . .

(c) Disclosure—Section 1421(b) of the Safe Drinking Water Act (42 U.S.C. 300h(b)) is amended by adding at the end the following:

‘(4) DISCLOSURES OF CHEMICAL CONSTITUENTS—

‘(A) IN GENERAL—A person conducting hydraulic fracturing operations shall disclose to the State (or to the Administrator, in any case in which the Administrator has primary enforcement responsibility in a State), by not later than such deadlines as shall be established by the State (or the Administrator)—

‘(i) before the commencement of any hydraulic fracturing operations at any lease area or a portion of a lease area, a list of chemicals and proppants intended for use in any underground injection during the operations (including identification of the chemical constituents of mixtures, Chemical Abstracts Service numbers for each chemical and constituent, material safety data sheets if available, and the anticipated amount of each chemical to be used); and

“(ii) after the completion of hydraulic fracturing operations described in clause (i), the list of chemicals and proppants used in each underground injection during the operations (including identification of the chemical constituents of mixtures, Chemical Abstracts Service numbers for each chemical and constituent, material safety data sheets if available, and the amount of each chemical used).

Source: Bill Text, 113th Congress (2013–2014), S.1135.IS. http://thomas.loc.gov/cgi-bin/query/z?c113:S.1135:. Accessed on May 25, 2014.

Powder River Basin Resource Council, Wyoming Outdoor Council, Earthworks, and Center for Effective Government (Formerly OMB Watch) vs. Wyoming Oil and Gas Conservation Commission and Halliburton Energy Services, Inc., 2014 WY 37 (2014)

Wyoming was the first state to require that companies using hydraulic fracking in their operations disclose to the public the chemicals used in that practice. In an early test of the law, the state Oil and Gas Conservation Commission agreed with the Halliburton Company that it was not bound by that law and that it was not required to disclose the chemicals it used in its fracking operations. The case reached the Wyoming Supreme Court in October 2013 and was decided in March 2014. The court returned the case to a lower court for a rehearing based on a technical problem with the earlier decision but also commented on the commission's decision not to enforce the state law.

[The court frames the question before it:]

Appellants present a single issue on appeal:

Whether the Supervisor of the Wyoming Oil and Gas Conservation Commission acted arbitrarily and unlawfully in denying Appellants' request for public records documenting the identities of chemicals used in hydraulic fracturing operations in the state.

Although varying in tone and tint, the Commission and Halliburton Energy Services, Inc. (Halliburton) restate essentially the same issue, asking us to apply the APA [Wyoming

Administrative Procedure Act]. However, the denial of Appellants' WPRA request does not fall under the APA. We therefore restate the issue to be determined as follows:

> Did the district court err in determining that the information sought in Appellants' public records request concerning the identity of certain chemicals used in hydraulic fracturing fluids are exempt from public disclosure as trade secrets under the WPRA [Wyoming Public Records Act]

[With respect to the substantive issue involved, the court said that the state had simply not provided the courts with sufficient information to justify withholding information about the fracking chemicals:]

The wisdom of the legislature's approach is apparent in this case. The record consists of information quite omnifarious, from irrelevant communications with various members of the public to trade secret protection requests from companies, the truth of which cannot be tested in a meaningful way. The record does not tell us what specific information the Supervisor relied upon, or why he did so, and therefore neither this Court nor the district court has a sufficient basis to determine whether he acted properly or not. On the other hand, as the WPRA and cases cited above point out, the district court is in a position to resolve issues of credibility, decide in the first instance whether the documents in question should be public or not, and create a meaningful record for appellate review by using the show-cause procedure in the WPRA.

[The court then remanded the case back to the district court with instructions as to how it was to proceed in deciding the case:]

Because the district court reviewed the Commission Supervisor's decision under the APA, we must reverse and remand. The district court is directed to determine whether it will permit Appellants to amend their existing pleadings to request and issue an order to the Supervisor to show cause as to why the documents requested should not be produced, or dismiss the case, which will permit Appellants to file a new action. In either event, the district court is required to determine as a matter of fact on evidence presented to it whether the information sought is a trade secret, and not whether the Supervisor acted arbitrarily or capriciously under the deferential administrative standards applied in the original proceedings. The district court will have to review the disputed information on a case-by-case, record-by-record, or perhaps even on an operator-by-operator basis, applying the definition of trade secrets set forth in this opinion and making particularized findings which independently explain the basis of its rulings in each.

Source: In the Supreme Court, State of Wyoming, 2014 WY 37. http://www.courts.state.wy.us/Opinions/2014WY37.pdf. Accessed on May 25, 2014.

Norse Energy Crop. vs. Town of Dryden et al. (2014)

One of the fundamental issues surrounding the use of fracking in the extraction of oil and gas is the question as to who has authority over land-use decisions in such operations. Many local authorities believe that they have such authority based on their historic rights of establishing zoning regulations in their municipalities. In many cases, state governments have rejected that notion and suggested that the production of gas and oil is of such importance to the state economy that licensing of gas and oil operations and related matters belongs in the realm of state authority and that that authority trumps local zoning ordinances and other attempts by cities, towns, counties, and other entities to regulate or restrict zoning. An important legal decision on this matter was issued in June 2014 when the highest court in the state of New York, the Court of Appeals, ruled in favor of two local municipalities, the towns of Dryden and Middlefield, against oil and gas companies and associated corporations. The crux of the court's decision is reprinted here.

Norse and CHC do not dispute that, absent a state legislative directive to the contrary, municipalities would ordinarily possess the home rule authority to restrict the use of land for oil and gas activities in furtherance of local interests. They claim, however, that the State Legislature has clearly expressed its intent to preempt zoning laws of local governments through the OGSML's "supersession clause," which reads:

> "The provisions of this article [i.e., the OGSML] shall supersede all local laws or ordinances relating to the regulation of the oil, gas and solution mining industries; but shall not supersede local government jurisdiction over local roads or the rights of local governments under the real property tax law" (ECL 23-0303 [2] [emphasis added]).

According to Norse and CHC, this provision should be interpreted broadly to reach zoning laws that restrict, or as presented here, prohibit oil and gas activities, including hydrofracking, within municipal boundaries.

. . .

[In a previous case, the Frew Run dispute], we held that this question may be answered by considering three factors: (1) the plain language of the supersession clause; (2) the statutory scheme as a whole; and (3) the relevant legislative history.

. . .

Nothing in the legislative history undermines our view that the supersession clause does not interfere with local zoning laws regulating the permissible and prohibited uses of municipal land. Indeed, the pertinent passages make no mention of zoning at all, much less evince an intent to take away local land use powers. Rather, the history of the OGSML and its predecessor makes clear that the State Legislature's primary concern was with preventing wasteful oil and gas practices and ensuring that the Department had the means to regulate the technical operations of the industry.

. . .

In sum, application of the three Frew Run factors—the plain language, statutory scheme and legislative history—to these appeals leads us to conclude that the Towns appropriately acted within their home rule authority in adopting the challenged zoning laws. We can find no legislative intent, much less a requisite "clear expression," requiring the preemption of local land use regulations.

. . .

Manifestly, Dryden and Middlefield engaged in a reasonable exercise of their zoning authority as contemplated in Gernatt when they adopted local laws clarifying that oil and gas extraction and production were not permissible uses in any zoning districts. The Towns both studied the issue and acted within their home rule powers in determining that gas drilling would permanently alter and adversely affect the deliberately-cultivated, small-town character of their communities. And contrary to the dissent's posture, there is no meaningful distinction between the zoning ordinance we upheld in Gernatt, which "eliminate[d] mining as a permitted use" in Sardinia (id. at 683), and the zoning laws here classifying oil and gas drilling as prohibited land uses in Dryden and Middlefield. Hence, Norse's and CHC's position that the town-wide nature of the hydrofracking bans rendered them unlawful is without merit, as are their remaining contentions.

Source: In the Matter of Mark S. Wallach . . . [New York Court of Appeals]. http://www.nycourts.gov/ctapps/Decisions/2014/Jun14/130-131opn14-Decision.pdf. Accessed on July 11, 2014.

Communication from the Commission to the European Parliament, the Council, the European Economic and Social Committee and the Committee of the Regions on the Exploration and Production of Hydrocarbons (Such as

Shale Gas) Using High Volume Hydraulic Fracturing in the EU (2014)

As of late 2014, there were no commercial fracking operations of shale oil or gas in the European Union. As this document indicates, a few pilot projects have been initiated and commercial production "could start in 2015–2017 in the most advanced member states. In anticipation of such an event, the European Commission has discussed possible political, social, economic, environmental, legal, and other issues that may arise as a result of fracking operations. They published the results of their studies as a set of Commission Recommendations on January 22, 2014, in the Official Journal of the European Union. *In a common practice, they also released a* Communication *explaining their recommendations in a somewhat condensed and simpler form. The core of those recommendations is as follows:*

In particular, the Recommendation invites Member States, when applying or adapting their legislation applicable to hydrocarbons involving high volume hydraulic fracturing, to ensure that:

a strategic environmental assessment is carried out prior to granting licenses for hydrocarbon exploration and/or production which are expected to lead to operations involving high-volume hydraulic fracturing in order to analyse and plan how to prevent, manage and mitigate cumulative impacts, possible conflicts with other uses of natural resources or the underground;

a site specific risk characterisation and assessment is carried out, related to both the underground and the surface, to determine whether an area is suitable for safe and secure exploration or production of hydrocarbons involving high volume hydraulic fracturing. It would inter alia identify risks of underground exposure pathways such as induced fractures, existing faults or abandoned wells;

baseline reporting (e.g. of water, air, seismicity) takes place, in order to provide a reference for subsequent monitoring or in case of an incident;

the public is informed of the composition of the fluid used for hydraulic fracturing on a well by well basis as well as on waste water composition, baseline data and monitoring results. This is needed to ensure that the authorities and the general public have factual information on potential risks and their sources. Improved transparency should also facilitate public acceptance;

the well is properly insulated from the surrounding geological formations, in particular to avoid contamination of groundwater;

venting (release of gases into the atmosphere) is limited to most exceptional operational safety cases, flaring (controlled burning of gases) is minimised, and gas is captured for its subsequent use (e.g. on-site or through pipelines). This is needed to mitigate negative effects of emissions on the climate, as well as on local air quality.

Source: "Communication from the Commission to the European Parliament, the Council, the European Economic and Social Committee and the Committee of the Regions on the exploration and production of hydrocarbons (such as shale gas) using high volume hydraulic fracturing in the EU." March 17, 2014. http://eur-lex.europa.eu/legal-content/EN/TXT/PDF/?uri=CELEX:52014DC0023R(01)&from=EN. Accessed on July 10, 2014. The Commission Recommendations are available online at http://eur-lex.europa.eu/legal-content/EN/TXT/PDF/?uri=CELEX:32014H0070&from=EN. Accessed on July 10, 2014.

OIL PIPE LINE

Research

Introduction

Fracking has become a topic of some interest and controversy in the last few decades, even though the procedure has been known for some time. Because of recent events, an increasingly large number of books, articles, Internet posts, and reports have been produced dealing with one or another phase of this topic. This chapter offers an annotated bibliography of a sample of these publications. In the relatively limited space provided, this collection can be no more than suggestive of the wide range of books and articles available, ranging from broad introductions intended for the general public to very specific publications of greater technical sophistication. Some publications are available in both print and electronic form and are so indicated in the listing.

Books

Bamberger, Michelle, and Robert Oswald. 2014. *The Real Cost of Fracking: How America's Shale-Gas Boom Is Threatening Our Families, Pets, and Food.* Boston, MA: Beacon Press.

An operator walks by the main oil line on offshore oil drilling platform 'Gail' operated by Venoco, Inc. off the coast of Santa Barbara, California. (AP Photo/Chris Carlson)

The authors of this book are a pharmacologist and a veterinarian. They purport to explain how the fracking process is damaging the natural environment and posing a threat to the health of humans and animals who live in the area of fracking operations.

Brasch, Walter M. 2013. *Fracking Pennsylvania: Flirting with Disaster.* Carmichael, CA: Greeley & Stone.

This book is divided into three major sections: an introduction to historical, political, and economic issues surrounding fracking; health and environmental issues; and a selection of social issues. As the title suggests, the author centers much of his story about the procedure in the state of Pennsylvania, exploring the development of fracking technology there and the effects that technology has had on the state's economic, political, and social character.

Bryce, Robert. 2008. *Gusher of Lies: The Dangerous Delusions of "Energy Independence."* New York: Public Affairs.

The author provides a superb review of the recent history of oil and gas exploration and the excitement caused in the industry and in the U.S. government by the apparent possibility of achieving energy independence. He goes on to explain why he thinks such a line of thinking is pure fantasy and that neither is energy independence likely to occur nor would it be a good thing for the United States and the rest of the world were it to do so.

Committee on Induced Seismicity et al., National Research Council. 2013. *Induced Seismicity Potential in Energy Technologies.* Washington, DC: National Academy Press.

This publication was prepared by a committee of specialists in the area of induced seismic events. Its purpose is to review and evaluate existing information on the relationship between extraction of fluid fuels from the earth and seismic events. The book also takes note of existing gaps in knowledge about such events and areas in which there is need for further research.

Del Percio, Stephen, and J. Cullen Howe. 2014. "The Legal and Regulatory Landscape of Hydraulic Fracturing." New Providence, NJ: LexisNexis.

This publication is divided into five sections that cover a general background to fracking technology, federal legislation and regulations, state legislation and regulations, important litigation involving fracking, and special recommendations for landowners regarding the leasing and/or sale of their land for fracking operations.

Donaldson, Erle C., Waqi Alam, and Nasrin Begum. 2013. *Hydraulic Fracturing Explained: Evaluation, Implementation, and Challenges*. Houston, TX: Gulf Publishing Company.

This book provides an excellent general introduction to some of the technical aspects of the fracking process, with chapters on the evaluation of gas shale formations, mechanics of rock fracturing, fluids used in the fracking process, and environmental impacts of hydraulic fracturing.

Ferguson, Spencer, and Matthew T. Gilbert, eds. 2012. *Hydraulic Fracturing and Shale Gas Production: Issues, Proposals and Recommendations*. New York: Nova Science.

This book provides a review of current and proposed laws and regulations dealing with the disclosure of chemicals used in hydraulic fracturing at both the federal and state levels.

Gold, Russell. 2014. *The Boom: How Fracking Ignited the American Energy Revolution and Changed the World.* New York: Simon & Schuster.

The author provides an excellent historical review of the development of hydraulic fracturing, along with factors that have caused it to become a controversial practice in the United States and other parts of the world.

Graves, John H. 2012. *Fracking: America's Alternative Energy Revolution*. Ventura, CA: Safe Harbor International Publishing.

The author reviews energy issues facing the United States and the rest of the world and points out the approaching era that will be characterized by the end of fossil fuels and the development of new forms of alternative energy. He discusses the critical role that hydraulic fracturing is now playing and will continue to play during this period of transition.

Hayden, Tom. 2013. *Fracked in the Barnett Shale: Drilling for the Right Balance*, 2nd ed. Charleston, SC: T.L. Hayden.

The author says that the purpose of this book is to provide "a broad spectrum of the challenges that must be addressed" by individuals and communities who are faced with the prospect of dealing with fracking operations in their vicinity.

Heinberg, Richard. 2013. *Snake Oil: How Fracking's False Promise of Plenty Imperils Our Future*. Santa Rosa, CA: Post Carbon Institute.

Heinberg is a well-known spokesman for and prolific writer about peak energy and the purported coming end of the age of fossil fuels. In this book, he examines the claims made for extending that period of human history through the means of hydraulic fracturing and explains why those claims are of limited validity.

Hillstrom, Kevin. 2013. *Fracking*. Detroit, MI: Lucent Books.

This book is part of the company's "Hot Topics" series of important current issues. It provides a general background to the topic of fracking and then offers arguments pro and con for its use as a recovery technology.

Holloway, Michael D., and Oliver Rudd. 2013. *Fracking: The Operations and Environmental Consequences of Hydraulic Fracturing*. Hoboken, NJ: John Wiley & Sons, Inc.; Salem, MA: Scrivener Publishing.

This book provides a very thorough introduction to the history and technology of hydraulic fracturing before turning to a consideration of the arguments pro and con that have been offered with regard to use of the process in the extraction of fossil fuels.

Kolb, Robert W. 2013. *The Natural Gas Revolution: At the Pivot of the World's Energy Future.* Upper Saddle River, NJ: Pearson Education.

The author refers to the discovery of shale gas and methods for extraction as the "revolution" of the title of this work. He begins with a detailed and comprehensive analysis of natural gas resources available in the United States, along with a review of the technologies involved in extracting, transporting, and storing natural gas. He also reviews the status of natural gas resources in other parts of the world.

Lane, C. Alexia. 2013. *On Fracking.* Victoria, BC: Rocky Mountain Books.

This book provides a general introduction to the topic of fracking, with special attention to its future role in dealing with the world's energy problems.

Levant, Ezra. 2014. *Groundswell: The Case for Fracking.* Toronto: McClelland & Stewart.

Levant explains why fracking is such a critical technology for dealing with the world's current energy problems and discusses and attempts to refute objections raised by opponents of the technology. He says that, in any case, the advantages provided by fracking far outweigh any disadvantages they may have.

Levi, Michael. 2013. *The Power Surge: Energy, Opportunity, and the Battle for America's Future.* Oxford, UK: Oxford University Press.

The author points out that the dispute over fracking often involves parties with very strong feelings about the subject that are sometimes inclined to offer arguments with somewhat questionable validity. He discusses the significance of the fracking process to America's and the world's current energy crisis, and tries to sort out the arguments that are most relevant and accurate in this ongoing debate.

Lustgarten, Abrahm. 2011. *Hydrofracked?: One Man's Mystery Leads to a Backlash against Natural Gas Drilling*. New York: ProPublica. Available at no cost from Amazon Digital Services.

This book tells the story of the actions taken by a Wyoming rancher after he discovered that the water on his land had begun to change color and produce a strange and unattractive odor, a change that he eventually attributed to hydrofacturing operations that were taking place near to his property.

Manning, Paddy. 2013. *What the Frack?: Everything You Need to Know about Coal Seam Gas*. Sydney, NSW: NewSouth Publishing.

This book focuses on the history and current status of hydraulic fracturing in Australia. The process was introduced with the promise of an important new energy resources with many environmental benefits. But, the author suggests, the reality of the experience has been somewhat different from that rosy promise.

McBroom, Matthew W. 2014. *The Effects of Induced Hydraulic Fracturing on the Environment: Commercial Demands vs. Water, Wildlife, and Human Ecosystems*. Toronto: Apple Academic Press.

This book is divided into three sections, the first of which deals with the effects of fracking on water systems; the second, on the effects on wildlife and the ecosystem; and the third, on human health.

Murphy, Adam R., ed. 2013. *Hydraulic Fracturing: Legal Issues and Relevant Laws*. New York: Nova Publishers.

The five chapters in this book deal with fracking operations vis-à-vis the Safe Drinking Water Act and Clean Water Act, as well as chemical disclosure issues and other selected legal issues.

Musialski, Cécile et al., eds. 2013. *Shale Gas in Europe: A Multidisciplinary Analysis with a Focus on European Specificities.* Deventer; Leuven: Claeys & Casteels Publishers.

This book consists of a number of papers that assess the status of shale gas extraction and development in the European Union. Comparisons of the progress of the technology in the EU are made with that in the United States, with lessons that can be learned from both sides.

Nabelhout, Ryan. 2014. *Fracking.* New York: Gareth Stevens Publishing.

This book is intended for a youth audience. It provides a general and simplified introduction to the technology and practice of fracking, along with a review of the issues associated with that technology.

Powers, Erica Levine, and Beth Kinne, eds. 2013. *Beyond the Fracking Wars: A Guide for Lawyers, Public Officials, Planners, and Citizens.* Chicago: American Bar Association. Section of State and Local Government Law.

Publishers of this book claim that it "avoids 'pro' and 'con' positions on the issue of fracking and provides basic information that allows individuals at all levels to learn more about the topic and make informed decisions about issues arising out of its use."

Prud'homme, Alex. 2013. *Hydrofracking: What Everyone Needs to Know.* Oxford, UK: Oxford University Press.

This book consists of three parts, the first of which provides a comprehensive review of fossil fuels and the nature of the hydraulic fracturing process. Part 2 discusses the elements of the controversy surrounding fracking, while

Part 3 focuses on the probable future of fracking in the United States and elsewhere.

Rao, Vikram. 2012. *Shale Gas: The Promise and the Peril.* Research Triangle Park, NC: RTI Press.

This book deals with the more general subject of shale gas, a resource for which hydrofracturing has become a critical method of recovery. The book is intended for a general audience and presents a good review of the environmental and human health issues involved in the exploration for and extraction of natural gas by fracking and other means.

Schultz, Aarik. 2012. *Hydraulic Fracturing and Natural Gas Drilling: Questions and Concerns.* New York: Nova Science Publishers.

After an introductory section on the mechanics of fracking, this book focuses on some political, social, economic, and other issues related to the use of the process in natural gas extraction, issues dealing with risks to the environment and human health, as well as problems associated with the Safe Clean Drinking Water Act.

Smith, Michael Berry, and Carl T. Montgomery. 2015. *Hydraulic Fracturing.* Boca Raton, FL: CRC Press.

The authors say that this book "busts the myths associated with hydraulic fracturing." They propose to outline a technology that will provide a safe, efficient, and effective process for removing maximum amounts of oil and gas from underground formations.

Spellman, Frank R. 2013. *Environmental Impacts of Hydraulic Fracturing.* Boca Raton, FL: CRC Press.

This book provides a very complete description of the fracking process and of its potential environmental consequences. It contains chapters on topics such as the nature of shale gas and its reservoirs, shale gas plays and other sources, the process of hydraulic fracturing and chemicals used in the process, possible environmental effects of fracking, and current laws and regulations dealing with the process.

Squire, Ann. 2013. *Hydrofracking: The Process That Has Changed America's Energy Needs*. New York: Children's Press.

This book is intended for young adults. It provides a general introduction to the topic of hydraulic fracturing and some of the issues involved with its use.

Thompson, Tamara. 2013. *Fracking*. Detroit, MI: Greenhaven Press.

This book is aimed at young adults as part of the company's "At Issue" series. It provides a general background to the subject with arguments both pro and con for the use of fracking in the extraction of fossil fuels.

Tootill, Alan. 2013. *Fracking the UK*. Charleston, SC: CreateSpace.

The author begins his book with a review of the fracking experience in the United States, before turning to the history and current status of the process in the United Kingdom

Wilber, Tom. 2012. *Under the Surface: Fracking, Fortunes and the Fate of the Marcellus Shale*. Ithaca, NY: Cornell University Press.

This book focuses on an area in the eastern United States that is especially rich in the fossil fuel products that can be obtained from hydraulic fracturing. It discusses the history and natural assets of the area along with the issues that have arisen as fracking in the area has become a major activity in a region that can benefit from the economic advantages provided by the technology.

Zuckerman, Gregory. 2013. *The Frackers: The Outrageous Inside Story of the New Billionaire Wildcatters*. New York: Portfolio Penguin.

This book tells the story of the development of hydraulic fracturing through biographical sketches of some of the most important individuals involved in the discovery, development, and implementation of the technology.

Articles

Adushkin, Vitaly V. et al. 2000, Summer. "Seismicity in the Oil Field." *Oilfield Review*. 12(Part 2): 2–17, http://www.earthworksaction.org/files/publications/FACTSHEET-Fracking Earthquakes.pdf. Accessed on July 10, 2014.

This article is a technical analysis of seismic events that may have been associated with oil and gas drilling activities in various parts of the former Soviet Union. Authors are members of the Institute of Dynamics of Geospheres of the Russian Academy of Sciences and the Ministry of Fuel and Energy of the Russian Federation.

Bligh, Shawna, and Chris Wendelbo. 2013. "Hydraulic Fracturing: Drilling Into the Issue." *Natural Resources and Environment* 27(3): 7–12.

This article explores federal and state laws and regulations relating to the process of hydraulic fracturing.

Boudet, Hilary et al. 2014. "'Fracking' Controversy and Communication: Using National Survey Data to Understand Public Perceptions of Hydraulic Fracturing." *Energy Policy* 65: 57–67.

The authors discover that the majority of Americans have heard little or nothing about the fracking controversy and do not know if they support or oppose the practice. Among those who have formed an opinion, about half support and half oppose hydraulic fracturing in the United States.

Chen, Jiangang et al. 2014. "Hydraulic Fracturing: Paving the Way for a Sustainable Future?" *Journal of Environmental and Public Health*, 1–10.

The authors note that fracking has made the United States the world's largest producer of natural gas, and that the process holds great potential for solving many of the world's energy challenges in the near future. They point out, however, that prospects for the process would be greatly improved if companies were more transparent about the chemicals used in the fracking process. They call for greater regulations requiring transparency in this regard by federal and state governments and by the industry itself. They also call for the removal of some of the exceptions granted fracking companies from laws such as the Clean Water Act and the Resource Conservation and Recovery Act.

Coulter, Gerald R. 1973. "Hydraulic Fracturing—New Developments." *Gulf Coast Association of Geological Societies Transactions* 23: 47–53.

This article is primarily of historical interest, providing a relatively easy-to-read description of the process of fracking during its earliest stages of use in the United States.

Cournil, Christel. 2013. "Adoption of Legislation on Shale Gas in France: Hesitation and/or Progress?" *European Energy and Environmental Law Review* 22(4): 141–151.

The author provides a review of the development of legislation on fracking in France that ends with the adoption of a law preventing the use of fracking for petroleum and gas exploration and extraction. She explains why she thinks the law was "designed hastily and has failed to satisfy any of the stakeholders involved."

Davies, Richard et al. 2013. "Induced Seismicity and Hydraulic Fracturing for the Recovery of Hydrocarbons." *Marine and Petroleum Geology* 45(4): 171–185.

The authors review data on 198 earthquakes of 1.0 or greater intensity since 1929 to discover their causes. They find that hydraulic fracturing is "not an important mechanism" for such events, but that is likely to become more important in the future.

de Rijke, Kim. 2013. "Hydraulically Fractured: Unconventional Gas and Anthropology." *Anthropology Today* 29(2): 13–17.

The author takes a somewhat unusual approach to dealing with the social issues related to the development of fracturing processes, focusing on topics such as politics, discourses, risk, and knowledge.

Dobb, Edwin. 2013. "The New Oil Landscape." *National Geographic* 223(3): 28–59.

A spectacular overview of the process of hydraulic fracturing is presented in this article with the usual collection of vivid photographs typical of this magazine to illustrate the main features of the story.

Duggan-Haas, Don, Robert M. Ross, and Warren D. Allmon. 2013. *The Science beneath the Surface: A Very Short Guide to the Marcellus Shale*. Ithaca, NY: Paleontological Research Institution.

This book was developed for the purpose of providing consumers, especially those in New York state, with a background about the science involved in the Marcellus Shale Play, with special attention to issues that arise out of the hydraulic fracking that has occurred and may occur in the future for the purpose of extracting natural oil and gas in the region.

Esswein, Eric J. et al. 2013. "Occupational Exposures to Respirable Crystalline Silica during Hydraulic Fracturing." *Journal of Occupational and Environmental Hygiene* 10(7): 347–356.

The article reports on a study of 11 sites in five states where fracking processes were being used to determine the risk posed by frac sand to workers at the sites. Frac sand is a specialized form of quartz often used as a proppant in fracking operations. Researchers found the workers at all sites were being exposed to levels of frac sand greater than limits set by the National Institute for Occupational Safety and Health (NIOSH), sometimes by as much as 10 times the recommended limit.

Fryzek, Jon et al. 2013. "Childhood Cancer Incidence in Pennsylvania Counties in Relation to Living in Counties with Hydraulic Fracturing Sites." *Journal of Occupational and Environmental Medicine* 55(7): 796–801.

Researchers explored the number of cancer cases among children living in two regions of Pennsylvania, some in counties where fracking was being conducted, and some in counties where there was no fracking. They found no significant differences in cancer rates between the two groups; results, they said, that "offer comfort concerning health effects of HF on childhood cancers."

Goldstein, Bernard D. et al. 2014. "The Role of Toxicological Science in Meeting the Challenges and Opportunities of Hydraulic Fracturing." *Toxicological Sciences* 139(2): 271–283.

Many of the substances used in fracking may have toxic effects on humans at every stage of the process, including transport of the materials to the fracking site, processes conducted at the site during the fracking process, and during the process of removing waste materials and other substances from the fracking location. This article provides an overview of what some of the toxicological issues might be at all stages of this process.

Gordalla, Birgit C., Ulrich Ewers, and Fritz H. Frimmel. 2013. "Hydraulic Fracturing: A Toxicological Threat for Groundwater and Drinking Water?" *Environmental Earth Sciences* 70(8): 3875–3893.

The authors report on their findings with regard to the degree of contamination of groundwater at three sites in Germany where fracking is being conducted. They recommend some possible approaches to regulation that will ensure the quality of backflow water from such wells.

Gross, Sherilyn A. et al. 2013. "Analysis of BTEX Groundwater Concentrations from Surface Spills Associated with Hydraulic Fracturing Operations." *Journal of the Air & Waste Management Association* 63(4): 424–432.

Researchers analyzed surface spills of chemicals used in hydraulic fracturing operations to determine the concentration of four hazardous chemicals, benzene, toluene, ethylbenzene, and xylene (BTEX) contaminating groundwater in the area of operations. They found some contamination in all areas studied with concentrations of BTEX chemicals ranging from 8 to 90 percent greater than National Drinking Water recommended concentrations. They also noted that directed efforts at dealing with this problem by well operators universally resulted in reduced levels of contaminant at the well sites.

Haapaniemi, Peter. 2013. "Feature Beyond the Headlines: While Coverage of Hydraulic Fracturing Has Focused on the Energy, There Are Side Benefits to Consider, among Them Jobs and Resurgent Manufacturing." *American Gas* 95(3): 28–31.

The author of this article discusses a number of potential benefits from the development and expansion of hydraulic fracturing operations for the recovery of natural gas.

Heikkila, Tanya et al. 2014. "Understanding a Period of Policy Change: The Case of Hydraulic Fracturing Disclosure Policy in Colorado." *Review of Policy Research* 31(2): 65–87.

This article reports on a research studied conducted in Colorado after the state had adopted regulations requiring companies to disclose the names of chemical they use in the fracking process. Researchers explored the ways in which individuals and groups on differing sides of the fracking debate viewed the new regulations and the factors that influenced these views.

Heuer, Mark A., and Zui Chuh Lee. 2014. "Marcellus Shale Development and the Susquehanna River: An Exploratory Analysis of Cross-Sector Attitudes on Natural Gas Hydraulic Fracturing." *Organization & Environment* 27(1): 25–42.

The authors report on a public opinion survey among stakeholders in fracking operations in the region described in the title on four issues: economic opportunity, protection of health and safety, preserving communities, and achieving energy security. They suggest that their data provide a baseline for future studies of a similar nature that can identify trends in public attitudes about fracking.

Holland, Austin A. 2013. "Earthquakes Triggered by Hydraulic Fracturing in South-Central Oklahoma." *Bulletin of the Seismological Society of America* 103(3): 1784–1792.

The author reports on a series of earthquakes that struck in south-central Oklahoma in January 2011 in an area where hydraulic fracturing was being conducted. A total of 116 quakes was identified with magnitude of 0.6 to 2.9.

"An In-Depth Look at Hydraulic Fracturing." 2013. *Canadian Mining Journal* 134(4): 40–41.

This article not only reviews fundamental information about fracking, but also provides background as to the history and current status of hydraulic fracturing in Canada.

Johnson, Corey, and Tim Boersma. 2013. "Energy (In)security in Poland the Case of Shale Gas." *Energy Policy* 53: 389–399.

The authors discuss the debate over hydraulic fracturing on both sides of the Atlantic Ocean and then consider the special case of fracking in Poland, a nation whose policymakers have been "among the most fervent" about fracking of any nation in Europe.

Kassotis, Christopher D. et al. 2014. "Estrogen and Androgen Receptor Activities of Hydraulic Fracturing Chemicals and

Surface and Ground Water in a Drilling-Dense Region." *Endocrinology* 155(3): 897–907.

The authors hypothesized that a number of chemicals used in hydraulic fracturing might have endocrine disrupting effects. In samples taken from four regions of Garfield County, Colorado, they found that 89 percent, 41 percent, 12 percent, and 46 percent of the chemicals identified could be so categorized.

Kiviat, Erik. 2013. "Risks to Biodiversity from Hydraulic Fracturing for Natural Gas in the Marcellus and Utica Shales." *Annals of the New York Academy of Sciences* 1286(1): 1–14.

The author argues that concerns about the effects of fracking on human health have been discussed at some length, but little attention has been paid to comparable risks to other organisms, such as brook trout, freshwater mussels, forest orchids, and salamanders. He suggests that these threats may place at least some of these species at risk for extinction.

Korfmacher, Katrina Smith et al. 2013. "Public Health and High Volume Hydraulic Fracturing." *New Solutions* 23(1): 13–31.

The authors note the increase in high-pressure hydraulic fracturing and point out that public health problems in a number of areas are likely to result from this procedure, including deterioration of surface and ground water quality, air quality, quality of life, worker health, sand mining and transport, and climate change.

Kotzé, Petro. 2013. "Hydraulic Fracturing: Adding to the Debate." *The Water Wheel* 12: 11–15.

This article provides an introduction to plans for extending the use of fracking in South Africa, with a review of known issues associated with the technology, especially with regard to its use in the nation.

Liroff, Richard A. 2012. "R&D Grand Challenges: A Roadmap for Addressing Environmental and Social Issues Associated with Horizontal Drilling and Hydraulic Fracturing." *Journal of Petroleum Technology* 64(7): 60–63.

The author notes that fracking holds enormous potential for the petroleum industry as well as to society at large in coming year, but the potential contribution of the process is restricted to some extent by a number of environmental and social issues, for which he has some possible solutions.

Malin, Stephanie. 2014. "There's No Real Choice but to Sign: Neoliberalization and Normalization of Hydraulic Fracturing on Pennsylvania Farmland." *Journal of Environmental Studies and Sciences* 4(1): 17–27.

The author reports on interviews with farmers in Bradford and Washington counties, Pennsylvania, and explains how these individuals incorporate environmental and liberal concerns about fracking into the decision-making process that almost inevitably results in their signing leases for use of their land for fracking operations.

Manning, Richard, and Danny Wilcox Frazier. 2013. "Bakken Business: The Price of North Dakota's Fracking Boom." *Harper's* 1954: 29–37.

This article explores the advantages and disadvantages that have accrued to the state of North Dakota as a consequence of the explosive growth of fracking in the state over the past few decades.

Maule, Alexis L. et al. 2013. "Disclosure of Hydraulic Fracturing Fluid Chemical Additives: Analysis of Regulations." *New Solutions* 23(1): 167–187.

The authors explore state and federal regulations concerning hydraulic fracturing regulations and review the variety of exceptions granted to companies from many environmental regulations that would otherwise seem to apply.

Mohan, Murali et al. 2013. "Microbial Community Changes in Hydraulic Fracturing Fluids and Produced Water from Shale Gas Extraction." *Environmental Science & Technology* 47 (22): 13141–13150.

The authors report on a study of the composition of bacterial communities in water used in fracking operations before and after they have passed through the fracking system. They found a significant difference in the bacterial ecology in pre- and posttreatment water, along with significant increases in mineral and radioactive content of released water, changes which may have "significant implications for disinfection as well as reuse of produced water in future fracturing operations."

Nicholson, Barclay R., and Stephen C. Dillard. 2013. "Analysis of Litigation Involving Shale and Hydraulic Fracturing." *International Energy Law Review* 2: 50–66; 3: 90–106. Also see http://www.nortonrosefulbright.com/files/us/images/publications/20130228WhitePaperShaleandHydraulicFracturing.pdf. Accessed on May 30, 2014.

The authors review and comment on a large number of cases regarding the use of hydraulic fracturing in locations around the world.

Nicot, John-Philippe et al. 2014. "Source and Fate of Hydraulic Fracturing Water in the Barnett Shale: A Historical Perspective." *Environmental Science & Technology* 48(4): 2464–2471.

This article focuses on the statistical characteristics of water involved in the hydraulic fracturing process in one widely used play, the Barnett Shale. Researchers tracked the source, use, and disposal of water over the years from 1981 to 2012 and reported on trends on various aspects of water use and disposal.

Nolen-Hoeksema, Richard. 2013. "Elements of Hydraulic Fracturing." *Oilfield Review* 25(2): 51–52.

This short article provides a very complete and well written explanation of the basic principles of hydraulic fracturing for the layperson. Also available online at http://www.slb.com/~/media/Files/resources/oilfield_review/ors13/sum13/defining_hydraulics.pdf. Accessed on May 26, 2014.

Penningroth, Stephen M. et al. 2013. "Community-Based Risk Assessment of Water Contamination from High-Volume Horizontal Hydraulic Fracturing." *New Solutions* 23(1): 137–166.

This article describes efforts to collect baseline information on waterways in regions where fracking operations are being or may in the future be carried out to determine the level of contaminants present as a result of the fracking. It also describes an interactive database that is being developed for this use.

Popkin, Jennifer H. et al. 2013. "Social Costs from Proximity to Hydraulic Fracturing in New York State." *Energy Policy* 62: 62–69.

The authors report on a study designed to find out how economic costs for residents in counties where fracking is taking place differ from those for residents in counties where such activities are not being conducted. They found out that electricity produced from fracking operations is much more expensive than electricity from conventional sources.

Rutqvist, Jonny et al. 2013. "Modeling of Fault Reactivation and Induced Seismicity during Hydraulic Fracturing of Shale-Gas Reservoirs." *Journal of Petroleum Science and Engineering* 107: 31–44.

Some concern has been expressed about the possibility that hydraulic fracturing operations may tend to increase the frequency and severity of earthquakes in adjacent regions. This study is one of a number of such research efforts attempting to quantify that risk. These authors claim that fracking is likely to give rise to only a small number of microquake events.

Saba, Tarek. 2013. "Evaluating Claims of Groundwater Contamination from Hydraulic Fracturing." *Oil and Gas Journal* 111(7): 80–89.

The author describes chemical and physical tests that should be conducted in determining whether groundwater has been contaminated by the presence of fracked wells in a region.

Schick, Robert M. et al. 2013. "Litigation Environment for Drilling and Hydraulic Fracturing." *Environmental Law Reporter* 43(3): 10221–10231.

Although most legal articles have dealt with laws and regulations relating to hydraulic fracturing, a lot of activity is currently under way in the field of litigation. This article summarizes the results of a discussion held in January 2013 among a variety of stakeholders involved with both plaintiffs and defendants in legal cases relating to fracking.

Selley, Richard C. 2012. "UK Shale Gas: The Story So Far." *Marine and Petroleum Geology* 31: 100–109.

This article provides a very nice overview of the development of shale gas fracking in the United Kingdom since the first experimental well was drilled in 1875.

Simon, John A. 2014. "Editor's Perspective—An Update on the Hydraulic Fracturing Groundwater Contamination Debate." *Remediation* 24(2): 1–9.

The author notes how intense the debate over hydraulic fracturing has become and explores some of the factors that may have resulted in this situation.

Stringfellow, William T. et al. 2014. "Physical, Chemical, and Biological Characteristics of Compounds Used in Hydraulic Fracturing." *Journal of Hazardous Materials* 275(30): 37–54.

The authors describe the properties of 81 chemicals used in the fracking process, for which toxicity data are not available for 30.

Tallent, Joshua M. 2013. "The Empire Fracks Back: The Case for Hydraulic Fracturing Strict Liability in New York." *Environmental Claims Journal* 25(1): 43–49.

The author suggests that fracking is going to become an increasingly popular procedure in New York State, resulting in an ever increasing number of lawsuits based on supposed environmental damage. He suggests that courts apply the most stringent interpretations of the law in such cases in order to encourage companies to routinely make environmental considerations an integral part of their planning for fracking operations.

Tollefson, Jeff. 2013. "China Slow to Tap Shale-Gas Bonanza." *Nature* 494(7437): 294.

The author explores some of the reasons that China has been slow to introduce fracking as a way of recovering its huge supplies of natural gas. He suggests that challenging geological problems and lack of infrastructure may be important contributing factors.

Walter, Laura. 2013. "What the Frack? Safety Concerns Surface in Hydraulic Fracturing." *EHS Today* 6(1): 31–34.

This article provides a nice overview of the nature of hydraulic fracturing and some of the environmental and health concerns associated with the technology.

Walton, John, and Arturo Woocay. 2013. "Environmental Issues Related to Enhanced Production of Natural Gas by Hydraulic Fracturing." *Journal of Green Building* 8(1): 62–71.

The authors list and describe a number of environmental issues that may develop in an area as the result of the use of fracking technology in well drilling.

Weinstein, Mark. 2013. "Hydraulic Fracturing in the United States and the European Union: Rethinking Regulation to Ensure the Protection of Water Resources." *Wisconsin International Law Journal* 30(4): 881–911.

The author begins with a long section explaining how fracking poses a serious threat to water resources. He then explores laws and regulations directed at fracking in both the United States and Europe, concluding with explanations of how each area can learn from the other in the development of adequate fracking regulation.

Weltman-Fahs, Maya, and Jason M. Taylor. 2013. "Hydraulic Fracturing and Brook Trout Habitat in the Marcellus Shale Region: Potential Impacts and Research Needs." *Fisheries* 38 (1): 4–15.

Fracking operations in the Marcellus Shale Play have been shown to pose a potential threat for a variety of plant and animal species. In this paper, the authors describe the types of threats faced by brook trout from hydrological, physical, and chemical factors.

Willie, Matt. 2011. "Hydraulic Fracturing and 'Spotty' Regulation: Why the Federal Government Should Let States Control Unconventional Onshore Drilling." Brigham Young University Law Review 5: 1743–1781.

The author points out the importance of fracking in the nation's future energy equation and suggests that federal regulations place an undue burden on companies that make use of the process. He claims that states are better equipped to devise and enforce the regulations needed for hydraulic fracking.

Reports

"The Arithmetic of Shale Gas," Yale Graduates in Energy Study Group, http://newsmanager.commpartners.com/cipammr/

downloads/Yale%20HF%20Economic%20Study.pdf. Accessed on July 10, 2014.

This study describes an effort to determine the relative economic benefits and costs of fracking, without regard to other issues, such as human health and environmental effects. The study concludes that "[e]conomic benefits, as estimated in as limited methodology as is reasonable, exceed costs to the community by 400-to-1."

Armendariz, Al. "Emissions from Natural Gas Production in the Barnett Shale Area and Opportunities for Cost-Effective Improvements," Environmental Defense Fund, January 26, 2009. http://www.edf.org/sites/default/files/9235_Barnett_Shale_Report.pdf. Accessed on July 19, 2014.

This report is one of the more technical studies of the effects of fracking on air and water contamination with an extensive review of the types of consequences the procedure may have for human health and the environment. An extended concluding section describes some methods by which these consequences can be ameliorated.

"Evaluation of Impacts to Underground Sources of Drinking Water by Hydraulic Fracturing of Coalbed Methane Reservoirs," U.S. Environmental Protection Agency, June 2004. Accessed as a zip file through http://water.epa.gov/type/groundwater/uic/class2/hydraulicfracturing/wells_coalbedmethanestudy.cfm. Accessed on July 9, 2014.

This report is one of the most famous documents in the history of the debate over fracking in that it concluded that the EPA had found no evidence in the study reported here that fracking has any harmful effects on drinking water quality. A number of questions were raised about the quality of the report, and the U.S. Congress later ordered the EPA to repeat the study.

"The Future of Natural Gas," An Interdisciplinary MIT Study, June 6, 2011. http://mitei.mit.edu/system/files/NaturalGas_Report.pdf. Accessed on July 9, 2014.

This report is one in the MIT series of "The Future of . . ." studies. It provides a superb review of the nature of natural gas, its probable future as a source of energy worldwide, and the technical and environmental aspects of its extraction, with heavy emphasis on the processes of hydraulic fracturing and directional drilling.

Goldman, Gretchen et al. "Towards An Evidence-Based Fracking Debate: Science, Democracy, and the Community Right to Know in Unconventional Oil and Gas Development," Union of Concerned Scientists, October 2013. http://www.ucsusa.org/assets/documents/center-for-science-and-democracy/fracking-report-full.pdf. Accessed on August 2, 2014.

After an introductory chapter on the status of fracking in the United States, this report reviews the scientific information currently available about fracking and related activities, discusses federal and state laws relating to fracking, advances arguments for the importance of the general public's knowledge about and understanding the nature of fracking and its consequences, and recommends policies and practices that are likely to lead to better decisions about the use of fracking by the petroleum industry.

Ground Water Protection Council and ALL Consulting. "Modern Shale Gas Development in the United States: A Primer." April 2009. http://energy.gov/sites/prod/files/2013/03/f0/ShaleGasPrimer_Online_4-2009.pdf. Accessed on July 9, 2014.

As the title of this report suggests, it is designed to provide an introduction to the topic of shale gas exploration and extraction with special attention on the role of hydraulic fracturing in those processes. The main sections of the report deal with the importance of shale gas, shale gas development in the United States, regulatory framework, and environmental considerations.

Hammer, Rebecca, Jeanne VanBriesen, and Larry Levine. "In Fracking's Wake: New Rules are Needed to Protect Our

Health and Environment from Contaminated Wastewater," National Resources Defense Council, May 2012. http://www.nrdc.org/energy/files/fracking-wastewater-fullreport.pdf. Accessed on July 8, 2014.

This report provides one of the most complete reports available on the character and potential effects of fracking wastewater and the new legislation and regulation that may be necessary to deal with the harmful effects of this waste material.

Hazen & Sawyer. "Impact Assessment of Natural Gas Production in the New York City Water Supply Watershed," New York City Department of Environmental Protection, December 2009. http://www.nyc.gov/html/dep/pdf/natural_gas_drilling/12_23_2009_final_assessment_report.pdf. Accessed on July 9, 2014.

This study was conducted to determine potential hazards posed to New York City's water supply system as a result of the growth of fracking operations in the Marcellus Play. The report consists of four major parts, the first of which deals with general issues about drinking water regulations and natural gas exploration and extraction. The three remaining sections focus on geology of the Marcellus shale area, growth characteristics of natural gas extraction in the area, and potential impacts of the technology on the city's drinking water.

Mall, Amy, Sharon Buccino, and Jeremy Nichols. "Drilling Down. Protecting Western Communities from the Health and Environmental Effects of Oil and Gas Production," Natural Resources Defense Council, October 2007. http://www.nrdc.org/land/use/down/down.pdf. Accessed on July 9, 2014.

This report starts with the premise that "oil and gas production is a dirty process" and proceeds to focus on the chemicals used in fracking that may pose a risk to human health and the environment and the air, water, and land pollution caused by fracking operations. The final chapter presents some simple and safe alternatives to the use of fracking.

Ridlington, Elizabeth, and John Rumpler. 2013. "Fracking by the Numbers: Key Impacts of Dirty Drilling at the State and National Level," Boston: Environment America. Available online at http://www.environmentamerica.org/sites/environment/files/reports/EA_FrackingNumbers_scrn.pdf. Accessed on May 26, 2014.

This report takes the position that hydraulic fracturing is a "highly polluting effort to unlock oil and gas" from sources in the United States. It describes the research conducted by the authors of this report from the Frontier Group and the Environment America Research & Policy Center, respectively in the areas of toxic wastewater, water use, chemical use, air pollution, land damage, and global warming emissions.

"Shale Gas: A Global Perspective," KPMG International, http://www.kpmg.com/Global/en/IssuesAndInsights/Articles Publications/Documents/shale-gas-global-perspective.pdf. Accessed on July 8, 2014.

The multinational oil and gas company KPMG provides this report on the availability, production, consumption, and related information on shale gas resources worldwide. An excellent overview of the topic.

"Study of the Potential Impacts of Hydraulic Fracturing on Drinking Water Resources. Progress Report," Washington, DC: U.S. Environmental Protection Agency, Office of Research and Development, December 2012. EPA/601/R-12/011, Table 11, page 29. Available online at http://www2.epa.gov/sites/production/files/documents/hf-report20121214.pdf#page=209. Accessed on May 26, 2014.

In 2010, the U.S. Congress directed the Environmental Protection Agency (EPA) to conduct a study on the relationship between hydraulic fracturing and the safety of drinking water in regions where the procedure was being used. In 2014, the EPA released a preliminary report describing the activities in which it was engaged to respond to this charge and the results it had obtained thus far. This 262-page report describes the EPA's work on this topic.

"Technically Recoverable Shale Oil and Shale Gas Resources: An Assessment of 137 Shale Formations in 41 Countries Outside the United States," U.S. Energy Information Administration, June 2013. http://www.eia.gov/analysis/studies/worldshalegas/pdf/fullreport.pdf. Accessed on May 31, 2014.

This report attempts to estimate the amount of risked resources (resources that are probably or possible sources of oil or gas) in 41 nations around the world and outside the United States. The five leading nations for risked gas resources are the United States, China, Argentina, Algeria, and Canada, and for risked oil resources, Russia, China, the United States, Argentina, and Libya.

"Testimony Submitted to the House Committee on Natural Resources, Subcommittee on Energy and Mineral Resources, Washington, D.C.," June 18, 2009. https://iogcc.publishpath.com/Websites/iogcc/Images/Additional-IOGCC-TestimonyJune2009.pdf. Accessed on July 9, 2014.

This report was prepared by the Interstate Oil and Gas Compact Commission (IOGCC) for presentation to the Subcommittee on Energy and Mineral Resources in its review of the effects of fracking on groundwater and other issues related to the practice. The report contains a review of the legal history surrounding the development of fracking, a survey of states' experience with fracking, state regulatory practices with regard to fracking, an IOGCC resolution on fracking and a collection of similar state resolutions on the practice, and a reprint of previous IOGCC testimony on fracking.

Internet

Behar, Michael. "Fracking's Latest Scandal? Earthquake Swarms," *Mother Jones*, April/May 2013. http://www.motherjones.com/environment/2013/03/does-fracking-cause-earthquakes-wastewater-dewatering?page=1. Accessed on July 10, 2014.

This highly respected muckraking magazine focuses on some of the most dramatic examples of seismic events that may have been caused by gas and oil extraction operations.

Biello, David. "Can Fracking Clean China's Air and Slow Climate Change?" Scientific American, January 27, 2014. http://www.scientificamerican.com/article/can-fracking-clean-chinas-air-and-slow-climate-change/. Accessed on July 11, 2014.

China is the "sleeping giant" in shale gas recovery in the world, with its enormous natural reserves largely unexploited to date. This article analyzes the possibility that fracking in China may help the nation produce a much larger quantity of relatively clean burning natural gas, which may help with both the nation's air pollution problems and global climate change.

"Chemicals in Natural Gas Operations," The Endocrine Disruption Exchange, http://endocrinedisruption.org/chemicals-in-natural-gas-operations/introduction. Accessed on July 10, 2014.

This web page reports on a study of the chemicals used in fracking and other drilling operations with regard to potential threat to the human endocrine system. It is part of a larger project on the general effect of low-dose and/or ambient exposure to chemicals that may interfere with the normal functioning of the human endocrine system.

Clauson, Doug. "Hydraulic Fracturing," ASTM International, http://www.astm.org/sn/features/hydraulic-fracturing-nd12.html. Accessed on May 30, 2014.

This article describes efforts by ASTM International (formerly the American Society for Testing and Materials) to develop new standards for fracking procedures define best practices in the industry for dealing with this relatively new technology.

Dahlen, Elizabeth. "Hydraulic Fracturing—Managing the Risk of Produced Water," Journal of Environmental Hydrology 21. http://www.hydroweb.com/protect/pubs/jeh/jeh2013/dahlen.pdf. Accessed on May 30, 2014.

This paper is written for water managers and asks how the process of hydraulic fracturing raises new problems about dealing with the water used in and produced by fracking processes.

Dearden, Lizzie. "Allow Fracking in National Parks, Says Outgoing Environment Agency Chief," The Independent, June 28, 2014. http://www.independent.co.uk/environment/allow-fracking-in-national-parks-says-outgoing-environment-agency-chief-9569919.html. Accessed on July 10, 2014.

The retiring chair of Great Britain's Environment Agency says that fracking is safe enough to use even in pristine natural environments like the nation's national parks. He claims that fears about the technology are "definitely exaggerated" if fracking is done properly.

Deutsch, Nicholas, and P. Randall Crump. "The Power Struggle over the Regulation of Hydraulic Fracturing," Industry Pulse, http://www.shb.com/newsevents/2014/ThePowerStruggleOvertheRegulationofHydraulicFracturing.pdf. Access on May 28, 2014.

The authors point out that regulation of activities such as fracturing has traditionally been the purview of state governments. But increasing attention to the possible environmental impacts of the process have caused more local governments to become involved in monitoring and controlling fracking in their areas. The authors analyze the elements involved in this controversy.

"DOE's Unconventional Gas Research Programs: 1976–1995," http://www.netl.doe.gov/kmd/cds/disk7/disk2/Final%20Report.pdf. Accessed on May 24, 2014.

Over a two decade period, the U.S. Department of Energy supported a number of research projects designed to test the viability of various types of technology designed to increase the production of natural gas from "unconventional" sources. Hydraulic fracturing was one of the technologies studied in some of these projects. This report provides an overview of the studies and their major results.

"Directional Drilling Technology," Environmental Protection Agency, http://www.epa.gov/cmop/docs/dir-drilling.pdf. Accessed on May 24, 2014.

This paper describes the technology of horizontal drilling, along with reports on the current status of the technology (as of about 2000) and research on the procedure.

"Earthquakes and Hydraulic Fracturing," Earthworks Fact Sheet, January 13, 2014. http://www.earthworksaction.org/files/publications/FACTSHEET-FrackingEarthquakes.pdf. Accessed on July 10, 2014.

This brief pamphlet provides basic information on the association—to the extent that there is any—between hydraulic fracturing and seismic events.

Fairley, Peter. "Desperate U.K. Turns to Shale Gas," MIT Technology Review, September 10, 2013. http://www.technologyreview.com/news/518936/desperate-uk-turns-to-shale-gas/. Accessed on July 10, 2014.

The author explains why the United Kingdom is now thinking so seriously about the use of fracking to take advantage of its natural gas resources. The web page is of considerable interest not only because of the article itself but also because of the number of cogent comments it inspired.

"Frac Sand," Geology.com. http://geology.com/articles/frac-sand/. Accessed on May 30, 2014.

Frac sand is a special form of highly purified quartz crystals widely used in fracking operations. This article provides detail about its properties and uses in the process.

"FracFocus," http://fracfocus.org/welcome. Accessed on May 24, 2014.

This interactive website was created by the Ground Water Protection Council and Interstate Oil and Gas Compact Commission to provide consumers with reliable information about the chemicals used in fracking operations in wells in their neighborhood.

"FracNews," http://www.frackusa.com/. Accessed on May 23, 2014.

This website has apparently been developed as an adjunct to John Graves's book on the topic, *Fracking: America's Alternative Energy Revolution* (see Books earlier). It focuses on what the author argues are the four key elements of fracking development and the controversy over the technology, namely the economy, the environment, jobs, and politics. The website also includes a blog that provides news about fracking along with commentaries about each of the four elements laid out by the author.

"Fracking: Gas Drilling's Environmental Threat," ProPublica. http://www.propublica.org/series/fracking. Accessed on July 10, 2014.

ProPublica is one of the most outspoken critics of the use of fracking for oil and gas extraction. This website contains six major sections on topics such as "The Story So Far," "Natural Gas Politics," and "Natural Gas Drilling: What We Don't Know." In addition, the site contains more than 160 additional articles on specific aspects of fracking, such as how best to measure health threats, "Chesapeake Energy's $5 Billion Shuffle," how gas and oil companies avoid paying royalties, EPA's retreat on previous fracking studies, and the threat posed by fracking in the Marcellus Play to Pennsylvania drinking water.

Hoffman, Joe. "Potential Health and Environmental Effects of Hydrofracking in the Williston Basin, Montana," Geology and Human Health, http://serc.carleton.edu/NAGTWorkshops/health/case_studies/hydrofracking_w.html. Accessed on July 10, 2014.

This excellent student presentation focuses on potential health issues resulting from fracking operations in a limited region of the Bakken gas and oil play in the upper Midwestern states.

"Hydraulic Fracturing," U.S. Geological Survey, http://energy.usgs.gov/OilGas/UnconventionalOilGas/HydraulicFracturing.aspx#3892235-overview. Accessed on May 26, 2014.

This website provides a comprehensive collection of information about hydraulic fracturing with a very nice collection of multimedia sources dealing with the topic, links to related Internet sites on fracking, and reference to a number of U.S. Geological Survey publications on the topic.

"Hydraulic Fracturing 101," Earthworks, http://www.earthworksaction.org/issues/detail/hydraulic_fracturing_101#.U67z-fldV8E. Accessed on June 28, 2014.

This website provides a superb general introduction to the subject of hydraulic fracturing along with an even-handed review of issues related to the practice. Some sections of special interest include those on groundwater contamination, air pollution, waste disposal, chemical disclosure, best practices in the field, and tips for landowners.

"Hydraulic Fracturing 101," Halliburton, http://www.halliburton.com/public/projects/pubsdata/Hydraulic_Fracturing/fracturing_101.html#. Accessed on May 24, 2014.

This web page provides a brief, but useful, introduction to fracking technology, with an excellent video showing the stages involved in the process. The page also has links to a number of other related sources of information.

"Is Fracking a Good Idea?" Debate Club. US News & World Report, http://www.usnews.com/debate-club/is-fracking-a-good-idea. Accessed on July 10, 2014.

The Debate Club provides an opportunity for the "sharpest minds" to exchange ideas on "the day's most important topics." This debate over fracking includes comments from a member of the U.S. House of Representatives, an associate professor of petroleum and geosystems engineering at the University of Texas at Austin, a member of the Marcellus Shale Protest group, and president of the Manhattan Borough in New York City. Each contributor writes a well-developed essay that usually includes useful links and references.

Global Frackdown. http://www.globalfrackdown.org/. Accessed on July 11, 2014.

Global Frackdown is an international conference first held in September 2012 with representatives from more than 20 countries who came together to protest the use of hydraulic fracturing for the extraction of oil and gas. The organization's most recent conference was held on October 11, 2014, at a number of sites around the world. This website provides a historical record of those meetings, a mission statement for the organization, a list of accomplishments in disrupting the use of fracking worldwide, guidance on carrying out protests against fracking, and information about natural gas drilling and fracking.

Kass, Stephen L. "Worldwide: Countries Approach Fracking with Interest and Caution," Mondaq, January 6, 2014. http://www.mondaq.com/unitedstates/x/284506/Climate+Change/Countries+Approach+Fracking+With+Interest+and+Caution. Accessed on July 11, 2014.

This article provides an excellent overview of the status of hydraulic fracturing in a number of countries around the world, with some thoughts on the possibility of growth of the practice in each country.

Linnitt, Carol. "Fracking the Future: How Unconventional Gas Threatens Our Water, Health and Climate," DeSmogBlog Society of British Columbia, 2010. http://www.desmogblog.com/fracking-the-future/desmog-fracking-the-future.pdf. Accessed on July 10, 2014.

This strongly worded booklet argues that the concept of natural gas as "clean energy" is a dangerous myth, that gas drilling poses a number of risks to human health and to the natural environment, that "Big Oil" has taken over the natural gas industry, and that companies involved in the extraction of gas use irresponsible and immoral political tactics. The booklet concludes with a number of recommendations for government and industry.

Luken, Christopher. "U.S. Surpasses Russia and Saudi Arabia as World's Leading Producer of Oil," North America's Shale Blog, July 10, 2014. http://www.northamericashaleblog.com/2014/07/10/u-s-surpasses-russia-and-saudi-arabia-as-worlds-leading-producer-of-oil/. Accessed on July 11, 2014.

This article describes the elements involved in the United States' achieving number one status among the world's oil producer, an accomplishment due primarily to the extensive use of fracking in oil and gas extraction operations.

McFeeley, Matthew. "State Hydraulic Fracturing Disclosure Rules and Enforcement: A Comparison," Natural Resources

Defense Council, http://www.nrdc.org/energy/files/Fracking-Disclosure-IB.pdf. Accessed on May 24, 2014.

Rules for the disclosure of chemicals used in hydraulic fracturing procedures vary widely from state to state in the United States. This document provides a comprehensive review of those differences, with a discussion of the types of regulations that are in force.

McMahon, Jeff. "Six Reasons Fracking Has Flopped Overseas." Forbes. April 7, 2013. http://www.forbes.com/sites/jeffmcmahon/2013/04/07/six-reasons-fracking-has-flopped-overseas/. Accessed on July 11, 2014.

McMahon attributes a failure to extend fracking to most countries other than the United States to price of the product, lack of an appropriate framework for the practice, poorly developed systems for dealing with property and mineral rights, insufficient infrastructure to support fracking, questions about water contamination and overuse, and lack of expertise in the field.

Morgan, James H. "Horizontal Drilling Applications of Petroleum Technologies for Environmental Purposes," Groundwater Monitoring & Remediation, http://info.ngwa.org/gwol/pdf/921856387.PDF. Accessed on May 24, 2014.

This journal article provides a brief history of horizontal drilling technology, along with its uses in protecting the environment during drilling operations.

Morgan, Rachel. "Fracking Wastewater Can Be Highly Radioactive," RSN, January 28, 2013. http://readersupportednews.org/news-section2/312-16/15754-fracking-wastewater-can-be-highly-radioactive. Accessed on July 10, 2014.

This article is the first of four in a series of radioactivity in fracking wastewater. A paid subscription is required for reading the final three articles in the series.

"Oil and Natural Gas Air Pollution Standards," Environmental Protection Agency, July 2, 2014. http://www.epa.gov/airquality/oilandgas/. Accessed on July 10, 2014.

In addition to recent regulatory actions by the EPA about air pollution resulting from hydraulic fracturing and related oil and gas activities, this website provides basic information about emissions caused by the oil and natural gas industry, technical information about the air toxics standards under review, public meetings held in connection with the rule-making process, and EPA's Natural Gas Star Program.

"Out of Control: Nova Scotia's Experience with Fracking for Shale Gas," Nova Scotia Fracking Resource and Action Coalition, April 2013. http://nofrac.files.wordpress.com/2013/04/out-of-control-full-report.pdf. Accessed on May 31, 2014.

Although only four fracked wells have been drilled in Nova Scotia, a number of environmental and other problems have persisted in the province. This report describes the history of fracking in Nova Scotia and the lessons that residents have learned about the procedure there.

"Overview of Petroleum Exploration & Production," http://www.mining.eng.chula.ac.th/CU_ETM/OverviewE&P.pdf. Accessed on July 8, 2014.

This book chapter provides one of the best general descriptions of the origin, exploration, and extraction of oil and gas available on the Internet. It is available from a number of different sources, this one being perhaps the most accessible for the reader.

"The Process of Shale Extraction," Energy from Shale, http://www.energyfromshale.org/hydraulic-fracturing. Accessed on July 10, 2014.

Energy from Shale is a production of the American Petroleum Institute, and so represents a particular viewpoint about shale gas and hydraulic fracturing. This website does contain some interesting written and visual information on various aspects of the fracking process and some social, economic, environmental, and other benefits offered by the procedure.

Rogers, S. Marvin. "History of Litigation Concerning Hydraulic Fracturing to Produce Coalbed Methane," http://iogcc.ok.gov/Websites/iogcc/Images/Marvin%20Rogers%20Paper%20of%20History%20of%20LEAF%20Case%20Jan.%202009.pdf. Accessed on May 24, 2014.

This paper reviews some of the most important legal actions taken between 1994 and 2009 related to the use of hydraulic fracturing for the extraction of natural gas from underground sources.

"Should the US Use Hydraulic Fracturing (Fracking) to Extract Natural Gas?" Alternative Energy Pros and Cons. ProCon.org, April 23, 2012. http://alternativeenergy.procon.org/view.answers.php?questionID=001732. Accessed on July 10, 2014.

The ProCon.org website provides essays and quotations in favor of and opposed to a variety of important current social issues. This page has statements from organizations such as the U.S. Department of Energy, the Wall Street Journal, the Ground Water Protection Council, and Earthworks, along with a number of individuals who write both for and against the use of fracking in oil and gas exploration and extraction.

State of Colorado Oil & Gas Commission. [untitled document], http://cogcc.state.co.us/library/GASLAND%20DOC.pdf. Accessed on May 24, 2014.

This document was prepared in response to the film Gasland about the purported effects of fracking on the natural environment and human health. The document is an effort, its authors say, "to correct several errors in the film's portrayal of the . . . incidents [that took place in Colorado]."

Strong, L. et al. "Biodegradation in Waters from Hydraulic Fracturing: Chemistry, Microbiology, and Engineering," *Journal of Environmental Engineering* volume 140, Special Issue: Environmental Impacts of Shale Gas Development, http://ascelibrary.org/doi/abs/10.1061/(ASCE)EE.1943-7870.0000792. Accessed on May 28, 2014.

The authors review the types of chemicals used in the fracking process, the potential environmental and human health impacts of these chemicals, and the potential for the use of biodegradation techniques for the removal of such chemicals.

Trembath, Alex et al. "Where the Shale Gas Revolution Came From: Government's Role in the Development of Hydraulic Fracturing in Shale," Breakthrough Institute Energy & Climate Program, http://thebreakthrough.org/blog/Where_the_Shale_Gas_Revolution_Came_From.pdf. Accessed on May 24, 2014.

This report provides a brief review of the history of the development of fracking operations in the United States, with special attention paid to the role of governmental agencies in the technology.

Truong, Alice. "The Pros and Cons of Shale Gas," How Stuff Works, http://science.howstuffworks.com/environmental/energy/pros-cons-shale-gas.htm. Accessed on July 10, 2014.

This five part series provides basic information about fracking along with discussions on some specific issues of importance, such as how fracking takes land from property owners, how deep fracking operations are conducted, and how gas is extracted from shale.

"US Surpasses Russia as World's Largest Natural Gas Producer," World Oil News Center, January 13, 2010. http://www.worldoil.com/US-surpasses-Russia-as-worlds-largest-natural-gas-producer.html. Accessed on July 11, 2014.

This news article discusses the attainment by the United States of the position of world's largest producer of natural gas, largely as a result of the use of fracking operations in gas extraction.

Uth, Hans-Joachim. "Technical Risks and Best Available Technology (BAT) of Hydraulic Fracturing in Unconventional Natural Gas Resources," *Environmental Earth Sciences*, http://link.springer.com/article/10.1007%2Fs12665-014-3126-8. Accessed on May 29, 2014.

The author reviews some of the risks associated with hydraulic fracking and lists some recommendations for practices that will reduce those risks. In general, he suggests that companies follow standard recommendations for the best available technology in fracking operations, whether they are required by law and regulations or not.

Vengosh, Avner et al. "A Critical Review of the Risks to Water Resources from Unconventional Shale Gas Development and Hydraulic Fracturing in the United States" Environmental Science & Technology, http://pubs.acs.org/doi/abs/10.1021/es405118y. Accessed on May 29, 2014.

The authors discuss four major sources of pollution to water resources from hydraulic fracturing: (1) the contamination of shallow aquifers with stray hydrocarbon gases; (2) the contamination of surface water and shallow groundwater from spills, leaks, and/or the disposal of inadequately treated shale gas wastewater; (3) the accumulation of toxic and radioactive elements in soil or stream sediments near disposal or spill sites; and (4) the overextraction of water resources for high-volume hydraulic fracturing that could result in water shortages or conflicts with other water users, especially in water-scarce areas.

"Water and Hydraulic Fracturing," American Water Works Association, http://www.awwa.org/Portals/0/files/legreg/documents/AWWAFrackingReport.pdf. Accessed on May 30, 2014.

This excellent white paper sets out the American Water Works Association's (AWWA) position on issues of water use in fracking operations. The publication provides an excellent general introduction to the general topic of hydraulic fracturing and outlines the issues related to water use associated with the procedure.

"Worker Exposure to Silica during Hydraulic Fracturing," Hazard Alert. OSHA-NIOSH, June 2012. https://www.osha.gov/dts/hazardalerts/hydraulic_frac_hazard_alert.pdf. Accessed on July 10, 2014.

The National Institute for Occupational Safety and Health (NIOSH) has identified exposure to airborne silica during the process of hydraulic fracturing as a health hazard to workers. This publication reviews the studies that led to this determination by the agency and discusses potential health hazards posed by silica in the air.

PRESSURE PIPE

Introduction

Hydraulic fracturing itself is a relatively new method for extracting fossil fuels from the Earth. However, the search for fossil fuel reserves and the variety of methods developed for their collection and transportation has gone on for many centuries. This chronology highlights some of the important events that have taken place over this period of time.

ca. 420 MYA (million years ago) Formation of Earth's petroleum and natural gas deposits began with the decay of microscopic plants and animals. This process continued intermittently over the next 400 million years.

1803 A resident of Baku named Gasimbey Mansurbeyov is credited with having dug the world's first oil in the Bibi-Heybat Bay, 18 and 30 meters of the coastline. He used only picks and shovels and no specialized equipment. The wells lasted only a short time as they were destroyed by inclement weather on the coast.

1821 The first natural gas well is drilled into Devonian shale near the city of Fredonia, New York. The gas is piped into the city, where it is used primarily for lighting.

Steve Lipsky demonstrates how his well water ignites when he puts a flame to the flowing well spigot outside his family's home in rural Parker County near Weatherford, Texas. (AP Photo/David J. Phillip)

1846 The first oil well to be drilled by mechanical means was sunk in the Bibi-Heybat suburb of Baku under the direction of Russian engineer F. N. Semyenov.

1857 An operation known as *fracture stimulation* is first used in Romania as a means of improving the porosity of oil- and gas-containing rock structures and, therefore, the efficiency of oil wells drilled in the region.

1858 Polish pharmacist Jan Józef Ignacy Łukasiewicz constructs the world's first refinery for the purpose of extracting kerosene from crude oil.

The first oil wells in North America are drilled under the supervision of Canadian entrepreneur James Miller Williams in southwestern Ontario. The wells were originally drilled to obtain drinking water but accidentally produced crude oil also.

1859 The first oil well in the United States (but not the first in the world nor the first in North America) is drilled near Titusville, Pennsylvania, by Colonel Edwin Drake. The well was, however, the first one drilled in North America with the purpose of extracting oil. The well was about 21 meters (69 feet) deep and produced up to 20 barrels of oil a day.

1865 Colonel Edward A. L. Roberts, a veteran of the Civil War, receives a patent for an "exploding torpedo" that can be used for increasing the flow of oil and gas from low-producing or played-out wells.

1875 Hydraulic fracturing was used for the first time in the United Kingdom to drill an experimental well near Battle, in East Sussex. The initial well reached a depth of about 300 meters before work was suspended. A second well was drilled to a depth of more than 600 meters before an explosion brought the work to a conclusion.

1891 The first patent for a horizontal drilling system is granted to John Smalley Campbell. Although Campbell's invention is designed for dental work, he pointed out ways in which the concept could be upgraded for use with oil wells also.

1896 California businessman Henry L. Williams oversees the drilling of the first commercially successful offshore petroleum wells near the town of Summerland, California.

1929 The first use of horizontal drilling for the extraction of oil is carried out at Texon, Texas. The technology was used as a way of draining oil out of a field, thus increasing its overall output.

The first patent for a multilateral well is issued to Texas petroleum engineer Leo Ranney. A multilateral well is one in which two or more horizontal wells are drilled from a single bore hole.

1939 American inventor Ira J. McCullough receives a patent for a method by which bullets can be fired through the lining in a bore hole to cause fracturing of rock surrounding a well, thus increasing the flow of oil and gas.

1947 The first modern, large-scale offshore drilling project is initiated about 10 miles offshore from Terrebonne Parish, Louisiana.

The first experimental hydraulic fracturing treatment in the United States was conducted in the Hugoton gas field in Grant County, Kansas.

1949 The first commercial fracturing procedures are carried out by the Halliburton company in Stephens County, Oklahoma, and Archer County, Texas.

1951 The Ghawar oil field in Saudi Arabia begins production of crude oil. The field is generally thought to be the largest oil producing site in the world.

1956 American geophysicist M. King Hubbert publishes a paper entitled "Nuclear Energy and the Fossil Fuels" that, for the first time, outlines the principles of peak oil, originating a "peak energy" movement that remains very influential in petroleum policy discussions to the present day.

1967 As part of its Project Plowshare program, designed to find peaceful uses of nuclear energy, the U.S. Atomic Energy Commission (AEC) carries out Project Gasbuggy, designed to test the ability of nuclear weapons to increase the flow of oil and gas from underground wells. Gasbuggy is later followed by two comparable projects, Project Rulison, in 1969, and Project Rio Blanco, in 1973.

1968 The Pan American Petroleum Company makes use of so-called massive hydraulic fracturing (or high-volume hydraulic fracturing) to recover natural gas from a well in Stephens County, Oklahoma. The process involves the use of a half million pounds of proppant pumped into the formation.

1969 The AEC carries out Project Rulison (see 1967) near Rulison, Colorado.

1970 This year marks the beginning of a decade in which the U.S. Congress passes a series of bills designed to protect human health and the natural environment from contamination by human activity, acts such as the National Environmental Policy Act of 1970, Clean Air Act of 1970, Occupational Safety and Health Act of 1970, Clean Water Act of 1972, Safe Drinking Water Act of 1974, Federal Land Policy and Management Act of 1976, Resource Conservation and Recovery Act of 1976, Surface Mining Control and Reclamation Act of 1977, and Comprehensive Environmental Response, Compensation and Liability Act of 1980.

1974 The AEC carries out Project Rio Blanco (see 1967) near Rifle, Colorado.

The U.S. Congress passes and President Gerald R. Ford signs the Safe Drinking Water Act. Possible contamination as a result of fracking is not mentioned in the act.

1976 The U.S. Department of Energy (DOE) initiates the Eastern Gas Shales Project, a joint research project of federal and state agencies and private companies to explore the possibility of increasing the amount of gas obtained from unconventional sources, such as hydraulic fracturing.

1980 The U.S. Congress passes the Windfall Profits Tax Act of 1980, which includes a section (the so-called Section 29) providing tax incentives for energy production from "unconventional" sources, such as fracking for natural gas. This provision remains in effect until 2002, when it is repealed.

1986 The Safe Drinking Water Act is updated and amended, but fracking is still excluded from any mention in the act.

DOE sponsors the drilling of a hole to extract natural gas from Devonian shales in Wayne County, West Virginia.

1989 The first use of horizontal drilling for the extraction of natural gas is conducted in the Barnett shale field near the Oklahoma–Texas border.

1992 The Supreme Court of Colorado rules on two issues related to hydraulic fracturing, in one of which they confirm a county's right to require special permits for the use of fracking operations in the county (*LaPlata County vs. Bowen/Edwards Association*) and in the other of which they reject the right of a city or town to totally ban the use of fracking within municipality limits (*Voss vs. Lundvall Brothers, Inc.*).

1994 The Legal Environmental Assistance Foundation (LEAF) petitions the Environmental Protection Agency (EPA) to withdraw its approval of the Alabama Underground Injection Control (UIC) program because it had not taken into consideration the effects of hydraulic fracturing on groundwater, as provided by the Safe Drinking Water Act. The EPA denies the LEAF petition in 1995. LEAF then files suit to force EPA to take its requested action. The court case continues until 2001 (q.v.).

1995 The EPA announces that it does not regulate, nor does it believe it has the authority to regulate, fracking of oil and gas wells.

1996 The Safe Drinking Water Act is amended again, still without any mention of the role of fracking in maintaining clean water sources.

The EPA issues a report suggesting that the natural gas industry in the United States is responsible for the release of about 2 percent of all methane that escapes into the atmosphere. The report remains the "gold standard" for this information for more than a decade and then is superceded by new reports that provide very different data on the topic.

1997 In the case of *Legal Environmental Assistance Foundation vs. U.S. EPA*, 118 F.3d 1467 (11th Cir. 1997), the 11th Circuit Court of Appeals rules that the injection of fluids for the purpose of conducting fracking must be regulated by the EPA under the Safe Drinking Water Act.

2001 In the case of *LEAF vs. EPA, State Oil and Gas Board of Alabama*, 276 F.3d 1253 (11th Cir. 2001), the 11th Circuit Court of Appeals rules that the state of Alabama UIC's program for regulating hydraulic fracturing meets the requirements of that act. The ruling brings to a close a series of court cases and disputes dating to 1994 (q.v.).

2003 The Range Resources company begins drilling an exploratory well in the Marcellus Shale region of Pennsylvania, the first such well in that formation. The Marcellus Shale has become one of the most heavily explored regions using fracturing in the United States.

2004 The EPA publishes an extensive study of the effects of hydraulic fracturing on the safety of drinking water. The agency concludes that hydraulic fracturing has "little or no effect" to underground sources of drinking water.

2005 The U.S. Congress passes the Energy Policy Act, which provides a specific and special exemption for hydraulic fracturing from meeting standards of the Safe Water Drinking Act. The exemption becomes known as the Halliburton Loophole after the company that apparently gains the most in its own fracking practices from the exemption.

2009 The Fracturing Responsibility and Awareness of Chemicals Act (FRAC Act) is introduced into the U.S. Congress by representatives Diana DeGette (D-CO), Maurice Hinchey (D-NY), and Jared Polis, (D-CO) in the House and by Bob Casey (D-PA) and Chuck Schumer (D-NY) in the Senate. The purpose of the act is to define hydraulic fracturing as a federally regulated activity under the Safe Drinking Water Act. It would require companies to disclose the chemicals that they use in the fracturing process.

2010 Wyoming becomes the first state to pass a law requiring companies to disclose the chemicals they use in fracking.

The film *Gasland* premiers. The film purports to describe the effect on the environment and human health in local communities where horizontal drilling and fracking are being practiced. A sequel to the film, *Gasland 2*, is released in 2013.

The United States becomes the world's largest producer of natural gas, an achievement that can be credited primarily to the expansion of hydraulic fracturing operations in the country.

2011 The French government adopts a law prohibiting the use of hydraulic fracturing for the exploration or extraction of petroleum products from shale formations. The law is the first of its kind in the world.

The U.S. Energy Information Administration (EIA) announces that the newly discovered Monterey Oil and Gas Play in California may contain as much as 15.4 billion gallons of tight oil. (But see **2014**.)

The EPA releases a study indicating that pollution of water wells near the town of Pavillion, Wyoming, have been contaminated by products from fracking operations taking place in the area. (But see **2013**.)

2012 Argentina signs an agreement with Chevron to permit exploration for and extraction of oil and gas in the vast reserves estimated for that country's petroleum assets. The agreement is one of the first such actions between national governments outside the United States and major multinational petroleum firms.

Bulgaria becomes the second country in Europe (after France) to ban fracking.

South Africa places a moratorium on fracking in its arid Karoo region and then decides to lift the moratorium and resume all drilling. The Department of Mineral Resources says that it is convinced by the arguments made by a number of energy companies that the operation is safe for human health and the natural environment.

Researchers at the Endocrine Disruption Exchange in Paonia, Colorado, call attention to a new class of compounds known collectively as *non-methane hydrocarbons* (NMHCs) in the effluent from fracked oil and gas wells with possible harmful effects on human health.

The Pennsylvania state legislature passes Act 13 which, among other provisions, prohibits cities, towns, counties, and other municipalities from regulating fracking operations within their jurisdiction. (But also see **2013** on this point.)

2013 The Supreme Court of Pennsylvania rules that portions of Act 13 are unconstitutional under the state constitution because they violate the Environmental Rights Amendment of the state constitution.

In the period between 2009 and 2013, the number of earthquakes of magnitude 3.0 or greater averages about 40 per year, compared to fewer than three per year in the preceding three decades.

Filmmakers Phelim McAleer and Ann McElhinney release a motion picture, *Frack Nation*, which they claim presents a very different view about fracking than that offered by the earlier film, *Gasland*. (See **2010**.)

Studies begin to appear suggesting that the amount of methane released during fracking operations may be many times higher than those first announced by the EPA in 1996 (q.v.)

The EPA reverses itself on a 2011 decision (q.v.) that fracking was responsible for contamination of freshwater wells near Pavillion, Wyoming. The agency announces that it would withdraw from further studies of that area and hand the project over to the state of Wyoming and the EnCana Corporation, which had been conducting drilling operations on the wells in question.

2014 The United States becomes the world's largest producer of oil, surpassing the former leader, Saudi Arabia. This achievement is credited largely to the increased use of hydraulic fracturing in the extraction of oil from shale.

The New York Court of Appeals, highest court in the state, rules that four towns in the state have the right to ban and/or regulate hydraulic fracturing as part of their zoning rights under the state constitution. Observers suggest that this could be a "death blow" for fracking in the state.

The Wyoming Supreme Court rules that a district court incorrectly granted exemptions to the Halliburton company from a state law requiring that it reveal the chemical it used in hydraulic fracturing procedures in the state. (Also see **2010**.)

The EIA revises its estimates of the amount of oil that may be extracted from the Monterey Play (see **2011**) to 600 million barrels, 96 percent less than its original estimate.

The Colorado Oil and Gas Conservation Commission orders High Sierra Water Services to stop disposing wastewater into one of its Weld County injection wells because of concerns about possible seismic events that may be associated with the practice.

As with most technical subjects, hydraulic fracturing is a field that makes use of a number of technical terms. The legal, economic, social, and political aspects of the subject also make use of terms with specialized meaning in the field. This chapter lists a number of the most common terms used when discussing fracking, although the list provided here is somewhat limited because of space. For more extensive glossaries on the terminology of hydraulic fracturing, also see the "Oil and Gas Well Drilling and Servicing Tool" (U.S. Department of Labor: https://www.osha.gov/SLTC/etools/oilandgas/glossary_of_terms/glossary_of_terms_s.html) and "Study of the Potential Impacts of Hydraulic Fracturing on Drinking Water Resources" (U.S. Environmental Protection Agency, Glossary: http://www2.epa.gov/sites/production/files/documents/hf-report20121214.pdf#page=209).

Acidize The process of adding an acid (usually hydrochloric acid) to fracturing fluid in order to increase the viscosity of the extracted fluid.

Additive Any substance that is combined with a base fluid (usually water) and proppant (usually sand) to create a fracturing fluid, which is then pumped into a rock formation.

Annulus The space between the casing and the borehole in a well.

API Well Number A unique number assigned to each individual well in the United States. An API Number consists of four parts, the state code, county code, well code, and wellbore code. API is an acronym for the American Petroleum Institute, which established the system.

Aquifer A water-bearing stratum of permeable rock, sand, or gravel.

Backflow The movement of liquids out of a well as the result of pressure in the wellbore. Backflow may consist of liquid hydrocarbons from the well, fracking fluids, produced water from the well, and other materials. Also referred to as flow back water or flow back fluid.

Base fluid A liquid or foam substance that is mixed with a proppant and to which additives are mixed or added to comprise a fracturing fluid. The most common base fluid is water. Also referred to as a **carrier fluid**.

Biocide A chemical that kills or disables microorganisms, such as bacteria. Using in fracking operations to eliminate the bacteria that are responsible for corrosion of piping and other equipment.

Borehole A hole drilled into the ground for the purpose of exploring for oil or natural gas or to extract those materials from the earth. Also see **wellbore**.

Breaker A chemical additive that reduces the viscosity of a fluid by breaking long-chain molecules into shorter segments.

Carrier fluid. See **base fluid**.

CAS An acronym for Chemical Abstract Service, a division of the American Chemical Society that finds, collects, organizes, and makes available all publicly disclosed information about chemical substances.

Cased hole A section of the wellbore in which casing and cement is installed.

Casing A pipe placed in a well for one of three purposes: (1) to prevent the wall of the hole from caving in, (2) to prevent movement of fluids from one geologic formation to another, or (3) to provide a means of maintaining control of formation fluids and pressure as the well is drilled.

Crosslinker A chemical that is added to a fracturing fluid to increase the viscosity of the fluid by catalyzing the combination of smaller molecules to make larger molecules.

Drainage radius The roughly circular area around a wellbore from which hydrocarbon flow into the wellbore.

Flowback water See **back flow**.

Fluid rate The velocity at which a gas or liquid flows into or out of a well. The units used are in volume per time period, commonly barrels per minute (BPM).

Frac sand A special form of high-purity quartz widely used as an additive in the hydraulic fracturing process.

Fracturing fluid A mixture used in hydraulic fracturing that consists of the base fluid (usually water), the proppant (usually sand), and one or more additives.

Gelling agent An additive that increases the viscosity of a fluid without affecting its other properties.

Groundwater Water that exists underground, usually in pores and crevices in the rock.

Hydraulic horsepower (HHP) A measure of pumping power in a fracking operated that is based on the pumping rate and pressure.

Methane An organic compound, a hydrocarbon, with the chemical formula CH_4.

Natural gas A gaseous fossil fuel that consists primarily of methane, often with smaller amounts of other substances, such as ethane, propane, butane, nitrogen, carbon dioxide, and hydrogen sulfide.

Non-methane hydrocarbons (NMHCs) A class of compounds found in the effluent from fracked oil and gas wells that may have harmful effects on human health. The class includes compounds such as ethane, propane, butane, ethene, and ethyne.

Perforations Holes drilled into the steel casing inserted into a wellbore that allow fracturing fluids to flow into the formation and later act as conduits through which hydrocarbons flow into the wellbore.

Play As in "petroleum play" or "natural gas play," a group of wells, oil fields, or prospective drilling sites characterized by a common geological description.

Produced water Water (usually salty) that originates from underground formations that escapes from a well in conjunction with oil or natural gas.

Proppant A finely divided solid material, such as sand or aluminum oxide, that is mixed with a fracturing fluid, usually water, to bring about the fracture of a rocky source of natural gas or oil.

Proppant concentration The amount of proppant per volume of fracking fluid. The most commonly used unit is pounds of proppant added per gallon of fracturing fluid (ppa).

Risked resources Oil and gas reserves are possible or probable sources of the resource that have not yet been developed. For contrast, see **unrisked resources**.

Shale A soft, finely stratified sedimentary rock formed from consolidated mud or clay. It can generally be split easily into fragile slabs.

Slurry A mixture of cement and water that is pumped into a wellbore to hold the casing in place and to provide a seal for the casing.

Surface injection pressure (SIP) The pressure at the surface of a well at which a fracturing fluid is being injected into a well. The most commonly used unit is pounds per square inch (psi).

Tight sands A sand or sandstone formation with low permeability.

Unrisked resources Sources of oil and gas that have been developed and from which the resource has already been produced.

Wellbore A hole drilled into a rock formation for the purpose of exploring for or extracting oil or natural gas. Also see **borehole**.

Acidization, 46
Acidize, 327
Act 13 (Pennsylvania), 109–110, 258–60, 324
Additive, 327
Air Pollution from fracking, statistics, 239–40
Al Falih, Khalid, 6
America's Natural Gas Alliance, 175–76
American Gas Association, 170–71
American Petroleum Institute, 172–74
Annulus, 327
API well number, 327
Aquifer, 328
Associated natural gas, 13

Backflow, 328
Bakken play, 49
Baku oil region, 32–33
Bannister, Thomas, 32
Base fluid, 328
Bazhenov Play, 17–18
Bell, Trudy E., 131–36
Berkshire Environmental Action Team, 220–23
Biocide, 328
Bitumen, 29, 31
Borehole, 328
Breaker, 328
Browner, Carol, 102
Bryce, Robert, 73–74
Burnett, H. Sterling, 65

Campbell, John Smalley, 50–51, 318
Carlyle, Ryan, 67
Cased hole, 328
Casey, Bob, 104
Casing, 328
Çelebi, Evliya, 32
Chayvo oil well, 20
Chemicals used in fracking, 236
Cheney, Richard, 103–4
Chesapeake Climate Action Network, 176–79
Christmas tree (petroleum drilling), 25
Coal, 3–6
  genesis, 7–12

Communication from the Commission to the European Parliament, the Council, the European Economic and Social Committee and the Committee of the Regions on the Exploration and Production of Hydrocarbons (Such as 270 Fracking Shale Gas) Using High Volume Hydraulic Fracturing in the EU, 270–72
Conductor hole, 24
Consumer Energy Alliance, 179–81
Crosslinker, 329
Crude oil, 13
Crude Oil Windfall Profit Tax Act of 1980, 243–46, 320–21

DeGette, Diana, 104
Della Valle, Pietro, 32
Desmarest, Thierry, 3
Directional drilling. *See* Horizontal drilling
Discovered unrecoverable petroleum reserves, 18
Dissolved natural gas, 13
Drainage radius, 329
Drake, Edwin, 33–34, 318
Drill string, 24–25
Drilling fluid, 26
Drilling mud. *See* Drilling fluid
Duckett, Jeffrey, 32

Eastern Gas Shales Project, 320
Eastman, H. John, 181–83
The Endocrine Disruption Exchange, 94, 226–28
Energy independence, xiv, 70–76
Energy Watch Group, 5
Engelder, Terry, 169, 184–85
Environment America, 185–87
Evaluation of Impacts to Underground Sources of Drinking Water by Hydraulic Fracturing of Coalbed Methane Reservoirs Study, 252–54
Everett, Bruce, 136–40

Farris, Riley "Floyd," 47–48
Fast, Robert, 47–48
Flaring, 25
Fluid rate, 329
Food & Water Watch, 187–90
Ford, Gerald R., 320
Fossil fuels, 3–4
  genesis, 7–12. *See also* Coal; Natural gas; Oil; Petroleum
FRAC Act of 2009. *See* Fracturing Responsibility and Awareness of Chemicals Act

Frack Nation (film), 324
Frac sand, 329
Fracking, xiv, 6–7
aesthetic issues, 100–101
air pollution, 92–96
direct economic benefits, 76–79
environmental benefits, 80–82
indirect economic benefits, 79–80
laws and regulations, 101
public opinion, 82–83
risks, 84
seismic events, 96
water contamination, 87–92
water use, 84–87
Fracking, number of wells in the United States, 238
Fractionating column. *See* Fractionating tower
Fractionating tower, 28
Fractionation (petroleum), 28–29
Fracture stimulation, 318
Fracturing fluid, 329
Fracturing Responsibility and Awareness of Chemicals Act, 104, 263–66, 322–23
Frasch, Herman, 45–46

Gasland (film), 323
Gasland 2 (film), 323
Gelling agent, 329
Geophones, 20
Gesner, Abraham, 33
Ghawar Oil Field, 14, 319
Gravitometers, 21
Ground Water Protection Council, 190–92
Groundwater, 329
Gusher, 23

Hagler, Gina, 140–45
Halliburton, Erle P., 192–94
Halliburton loophole, 103–4, 322
Halliburton Oil Well Cementing Company, 48, 266, 319
Hart, William, 194–97
Herodotus, 31
Hinchey, Maurice, 104
History of natural gas exploration, 39–43
History of petroleum exploration, 30–34
Horizontal drilling, xiv, 49–51
Hubbert, M. King, 3–4, 319
Hubbert curve, 3–5
Hughes, J. David, 66–67
Hydraulic fracturing, xiv, 6, 43
history, 44–49. *See also* Fracking
Hydraulic Fracturing Exclusions, 246–51
Hydraulic horsepower (HHP), 329
Hydrophones, 21

Independent Petroleum Association of America, 197–99
Ingraffea, Anthony R., 200–202
Intermediate hole, 24
International Energy Agency, 202–6
Interstate Natural Gas Association of America, 206–8
Interstate Oil and Gas Compact Commission, 208–10

Jaffe, Amy Myers, 66

Kermac 16, 40
Kerr McGee Oil Industries, 40
Kim, Won-Young, 99
Klare, Michael T., 74–75

La Plata County vs. Bowen/Edwards, Associates, Inc., 108–9, 321
LEAF vs. EPA, State Oil and Gas Board of Alabama, 322
Legal Environmental Assistance Foundation (LEAF), 102, 321
Legal Environmental Assistance Foundation vs. U.S. EPA, 251–53, 322
Levi, Michael, 75–76
Łukasiewicz, Ignacy, 33, 318

Mansurbeyov, Gasimbey, 317
Marcellus Play, 17, 196, 322
Marland, E. W., 211–13
McAleer, Phelim, 324
McClendon, Aubrey, 213–15
McCullough, Ira J., 47, 319
McElhinney, Ann, 324
Methane, 329
Mitchell, George, 49–50, 215–18
Mitchell Energy, 49, 51
Monterey Play, 65–66

National Association of Royalty Owners, The, 218–20
Natural gas, 329
  associated, 13
  dissolved, 13
  drilling technology, 23–26
  exploration, 20–23, 39–43
  imports (statistics), 70
  reserves, 242–43
  terminology, 13–19. *See also* Peak natural gas; Peak oil
Nixon, Richard, xiii
No Fracked Gas in Mass, 220–23
Non-methane hydrocarbons (NMHCs), 324, 329
Norse Energy Crop. vs. Town of Dryden et al., 268–70

Obama, Barack, xiii
Offshore oil/gas drilling, history, 32, 40–43
  statistics, 42–43

Oil, terminology, 13–19
Oil field, 13–14
  statistics, 14–16
Otto Cupler Torpedo Company, 45

Pastorkovich, Michael, 145–49
Peak coal, 4–5
Peak energy, 6
Peak natural gas. *See* Peak energy; Peak oil
Peak oil, 3, 63–68
Perforations, 330
Petroleum, 13
  consumption, 36, 37–39
  drilling technology, 23–26
  exploration, 20–23, 30
  import statistics, 70
  production statistics, 34–35, 68–70
  reserves, 68–70, 242–43
  terminology, 18–20
Petroleum chemistry, 27–30
Play (oil and/or natural gas), 14, 330
Polis, Jared, 104
Porter, David J., 89
Powder River Basin Resource Council, Wyoming Outdoor Council, Earthworks, and Center for Effective Government (Formerly OMB Watch) vs. Wyoming Oil and Gas Conservation Commission and Halliburton Energy Services, Inc., 266–68
Pressure parting, 47
Produced water, 330
Production hole, 24
Project Gasbuggy, 46, 319
Project Plowshare, 46, 319
Project Rio Blanco, 46, 320
Project Rulison, 46, 319
Proppant, 49, 330
Proppant concentration, 330
Pure Oil Company, 40

Ranney, Leo, 319
Remote sensing, 23
Risked petroleum reserves, 18, 240–42, 330
Roberts, Edward A. L., 44–45, 223–26, 318
Roberts' torpedo, 45
Robinson Township, et al. vs. Commonwealth of Massachusetts, 110, 260–63
Rock busting, 47
Rogers, Deborah, 75
Rühl, Christof, 67–68
Rumpler, John, 149–54

Safe Drinking Water Act, 101, 103, 236, 246, 263–64, 321, 322, 324
Semyenov, F. N., 32, 318
Shale, 330
Shale gas, 10–12
  reserves, 68–72, 237–42

Shale oil, 10–12
reserves, 68–72, 237–42
Slanted drilling. *See* Horizontal drilling
Slurry, 330
Stanolind Oil and Gas Company, 47–48, 192, 197
Starter hole. *See* Conductor hole
Straub, Lana, 154–58
Structural trap, 13
Struys, Jan, 32
Study of the Potential Impacts of Hydraulic Fracturing on Drinking Water Resources—Progress Report, 256–58
Superior Oil Company, 40
Surface injection pressure (SIP), 330

Technically recoverable petroleum reserve, 18, 240–42
Testimony Submitted to the House Committee on Natural Resources Subcommittee on Energy and Mineral Resources Washington, DC, June 18, 2009, Prepared by the Interstate Oil and Gas Compact Commission on Behalf of the Nation's Oil and Gas Producing States, 254–56
Tight sands, 330
TOXNET, 91

Unrisked resources, 331

Voss vs. Lundvall Brothers, Inc., 321

Walter, Laura, 158–62
Wanjek, Christopher, 162–66
Ward, Tom L., 228–30
Water use in fracking, statistics, 239
Waterkeeper Alliance, 230–33
Wellbore, 331
Williams, Henry L., 40, 319
Williams, James Miller, 318
Wilson, Weston, 103
Worstall, Tim, 66

About the Author

David E. Newton holds an associate's degree in science from Grand Rapids (Michigan) Junior College, a BA in chemistry (with high distinction) and an MA in education from the University of Michigan, and an EdD in science education from Harvard University. He is the author of more than 400 textbooks, encyclopedias, resource books, research manuals, laboratory manuals, trade books, and other educational materials. He taught mathematics, chemistry, and physical science in Grand Rapids, Michigan, for 13 years; was professor of chemistry and physics at Salem State College in Massachusetts for 15 years; and was adjunct professor in the College of Professional Studies at the University of San Francisco for 10 years.

Previous books for ABC-CLIO include *Global Warming* (1993), *Gay and Lesbian Rights—A Resource Handbook* (1994, 2009), *The Ozone Dilemma* (1995), *Violence and the Mass Media* (1996), *Environmental Justice* (1996, 2009), *Encyclopedia of Cryptology* (1997), *Social Issues in Science and Technology: An Encyclopedia* (1999), *DNA Technology* (2009), and *Sexual Health* (2010). Other recent books include *Physics: Oryx Frontiers of Science Series* (2000), *Sick!* (4 volumes; 2000), *Science, Technology, and Society: The Impact of Science in the 19th Century* (2 volumes; 2001), *Encyclopedia of Fire* (2002), *Molecular Nanotechnology: Oryx Frontiers of Science Series* (2002), *Encyclopedia of Water* (2003), *Encyclopedia of Air* (2004), *The New Chemistry* (6 volumes; 2007), *Nuclear Power* (2005), *Stem Cell Research* (2006), *Latinos in the Sciences, Math, and Professions* (2007), and *DNA Evidence and*

Forensic Science (2008). He has also been an updating and consulting editor on a number of books and reference works, including *Chemical Compounds* (2005), *Chemical Elements* (2006), *Encyclopedia of Endangered Species* (2006), *World of Mathematics* (2006), *World of Chemistry* (2006), *World of Health* (2006), *UXL Encyclopedia of Science* (2007), *Alternative Medicine* (2008), *Grzimek's Animal Life Encyclopedia* (2009), *Community Health* (2009), *Genetic Medicine* (2009), *The Gale Encyclopedia of Medicine* (2010–2011), *The Gale Encyclopedia of Alternative Medicine* (2013), *Discoveries in Modern Science: Exploration, Invention, and Technology* (2013–2014), and *Cengage Science in Context* (2013–2014).